AF845688

Palgrave Studies in European Union Politics

Series Editors
Michelle Egan, American University, Washington, USA
William E. Paterson, Aston University, Birmingham, UK
Kolja Raube, KU Leuven, Leuven, Belgium

Following on the sustained success of the acclaimed European Union Series, which essentially publishes research-based textbooks, Palgrave Studies in European Union Politics publishes cutting edge research-driven monographs. The remit of the series is broadly defined, both in terms of subject and academic discipline. All topics of significance concerning the nature and operation of the European Union potentially fall within the scope of the series. The series is multidisciplinary to reflect the growing importance of the EU as a political, economic and social phenomenon. To submit a proposal, please contact Senior Editor Ambra Finotello ambra.finotello@palgrave.com. This series is indexed by Scopus.

Editorial Board

Laurie Buonanno (SUNY Buffalo State, USA)
Kenneth Dyson (Cardiff University, UK)
Brigid Laffan (European University Institute, Italy)
Claudio Radaelli (University College London, UK)
Mark Rhinard (Stockholm University, Sweden)
Ariadna Ripoll Servent (University of Bamberg, Germany)
Frank Schimmelfennig (ETH Zurich, Switzerland)
Claudia Sternberg (University College London, UK)
Nathalie Tocci (Istituto Affari Internazionali, Italy)

George Dikaios

EU Climate Diplomacy Towards the IMO and ICAO

UNESCO
ChAIR
ON CLIMATE
DIPLOMACY
unesco

palgrave
macmillan

George Dikaios
National and Kapodistrian
University of Athens
Athens, Greece

ISSN 2662-5873 ISSN 2662-5881 (electronic)
Palgrave Studies in European Union Politics
ISBN 978-3-031-51122-6 ISBN 978-3-031-51123-3 (eBook)
https://doi.org/10.1007/978-3-031-51123-3

© The Editor(s) (if applicable) and The Author(s), under exclusive license to Springer Nature Switzerland AG 2024

This work is subject to copyright. All rights are solely and exclusively licensed by the Publisher, whether the whole or part of the material is concerned, specifically the rights of translation, reprinting, reuse of illustrations, recitation, broadcasting, reproduction on microfilms or in any other physical way, and transmission or information storage and retrieval, electronic adaptation, computer software, or by similar or dissimilar methodology now known or hereafter developed.
The use of general descriptive names, registered names, trademarks, service marks, etc. in this publication does not imply, even in the absence of a specific statement, that such names are exempt from the relevant protective laws and regulations and therefore free for general use.
The publisher, the authors, and the editors are safe to assume that the advice and information in this book are believed to be true and accurate at the date of publication. Neither the publisher nor the authors or the editors give a warranty, expressed or implied, with respect to the material contained herein or for any errors or omissions that may have been made. The publisher remains neutral with regard to jurisdictional claims in published maps and institutional affiliations.

Cover illustration: Magic Lens/Shutterstock

This Palgrave Macmillan imprint is published by the registered company Springer Nature Switzerland AG
The registered company address is: Gewerbestrasse 11, 6330 Cham, Switzerland

Paper in this product is recyclable.

'I am very convinced that without the EU we wouldn't have the *Initial Strategy on Reduction of Greenhouse Gas Emissions from Ships and the Carbon Offsetting and Reduction Scheme for International Aviation.* If Europe wouldn't have acted as one, [these two outputs would not have been adopted. Europe's position was strong] and the strength came from really acting with one voice, one strong voice'.

'IMO and ICAO instrumentalise this need of Europe to be progressive—not in a bad way. They want money to bridge the gaps and we are here to provide it—and I think that's fair. And, also, they want knowledge, and we provide capacity-building'.

Violeta Bulc

European Commissioner of Transport (2014–19)[1]

'Let's be honest. While [UN] member states have made some initial steps through the International Civil Aviation Organization and the International Maritime Organization to address emissions from shipping and aviation, current commitments are not aligned with the 1.5°C goal of the Paris Agreement. In fact, they are more consistent with warming way above 3°C'.

António Guterres

UN Secretary General (2017–today)[2]

[1] Interview no. 48

[2] The Guardian (2021).

Foreword

As a UNESCO Chairholder on Climate Diplomacy with particular interest and practice in the field, I am delighted to write the foreword of this book, which is the fruitful outcome of an original and complex doctoral research. Its originality stems from the fact that, despite the existence of an abundance of literature on European climate policy, no systematic empirical research had been conducted on its influence on policies in other international organizations. This is particularly evident in the ostensibly critical—in regard to the ongoing climate crisis—sectors of aviation and shipping which are producing high levels of pollution, with their combined carbon dioxide emissions being very high, and with no mitigation regulations in place. The latter are not included in the emissions of states, nor do they fall within the scope of the international climate negotiations under the United Nations Framework Convention on Climate Change (UNFCCC). Having developed the most ambitious climate policy so far, the European Union (EU) has been seeking to regulate these emissions for several years.

The book explores the said influence through the prism of climate diplomacy, a rather new addition to the vocabulary of politics and diplomacy, which frames climate change as an element of external action policy and highlights the need to integrate the climate objectives at the highest diplomatic level and across all policy areas.

The complexity of the research is highlighted through the scholarly occupation with a very technical, multifaceted, and interdisciplinary policy issue. The main tools for this endeavor stem from political science, however, George Dikaios masterfully employs processes and procedures that originate in legal thought (interpretation of regulatory framework, jurisdiction, competence, and compliance) and the political economy of international organizations (institutional effectiveness and internal working factors). The reader will find an analysis that builds on a wide array of source materials and evidence, as the author shows a remarkable ability to engulf both practice and theory into the intellectual framework of the book.

Dikaios takes us inside his thinking as he reflects on whether and to what extent the EU is influencing the international negotiations to reduce carbon dioxide emissions through the exercise of climate diplomacy. He provides a detailed and extended analysis of the degree of the EU's influence through diplomacy at two levels. On the one hand, he examines the internal processes of the EU's ability to influence its Member States to form a common position, which then will be supported in international negotiations. On the other hand, at the external level, he explores the EU's ability to directly influence international negotiations, through two case studies, namely the International Maritime Organization (IMO) and the International Civil Aviation Organization (ICAO) which perform regulatory action in shipping and aviation, respectively. Based on the theoretical framework of Normative Power Europe, the author provides a series of work assumptions which he then tests empirically. The fifty-four interviews conducted by the author with EU and EU Member State officials who have participated in international negotiations aiming to reduce emissions from aviation and shipping, illuminate opinions of key actors that cannot be illustrated by other sources (official documents or literature). These interviews showcase the different angles in their motives and preferences, as well as strategy tactics employed or deployed during the diplomatic process, both inside and outside the EU. The empirical testing helps the author draw comprehensive conclusions on the degree of the influence the EU has both on its Member States and the international organizations under study concerning the diffusion of European norms on climate targets and evaluate the practice of European climate diplomacy.

Dikaios shows that the EU's influence on emission reductions is greater in the ICAO than in the IMO where the corresponding domestic policy has not yet been fully crystallized. While the Member States are at the forefront of the negotiations, the EU's presence both in the preparatory stages and during the international negotiations is increasingly strengthened with the European Commission playing a leading role, trying to coordinate the Member States—sometimes even exerting suffocating pressure—and at the same time affecting third countries. This presence tends to intensify because of the implementation of the European Green Deal, which includes significant changes to emissions from ships and aircrafts and enhances the role of the Commission inside and outside the EU.

As Dikaios demonstrates, European climate diplomacy is founded on internal political processes that contribute to the diffusion of European norms. Institutional constraints (the scope of the EU competence internally and externally, its observer-only status in the two under study organizations) tend to be surpassed through the exercise of methodical and systematic diplomacy and an intensive and highly rigorous internal coordination process between the Commission and the Member States. However, as the author argues, although the EU pursues a normative climate diplomacy, it has so far failed to achieve substantial change in the sectors of aviation and shipping. EU's normative power is therefore incomplete. Systematic preparation, greater coordination, targeted strategy, and multilevel diplomacy that matures over time are important tools that can strengthen the EU's position in international negotiations.

Overall, the book covers an important angle, which has been neglected in the literature for this theoretical approach, of the connection between the internal and the external action of the EU. The author has married knowledge of detail with original conceptual thinking. The result is a book that is at once informative and challenging. It can help students, academics, and practitioners develop a shared vision and understanding of the European climate diplomacy as well as the challenges and opportunities in enhancing the EU's role in the international arena.

Prof. Emmanuella Doussis
UNESCO Chairholder on
Climate Diplomacy
National and Kapodistrian
University of Athens
Athens, Greece

Acknowledgments

This book is based on my doctoral thesis, which I defended at the Department of Political Science and Public Administration of the National and Kapodistrian University of Athens, under the supervision and mentoring of Professor Emmanuella Doussis, to whom I am very grateful and indebted for her academic and career guidance, collaboration, trust, and friendship. Further I cannot thank her enough for the totality of her contribution and for the honor she extended me by writing the Foreword of this book. It merits a mention, also, that is the first thesis written under the auspices of the UNESCO Chair on Climate Diplomacy, in which I have the great pleasure of currently being a Senior Researcher.

Additionally, the development of my thesis invaluably benefited from the insight, advice, and help of the other two members of my advisory committee (or co-promoters, as called in The Netherlands), Associate Professor Karolina Pomorska (Leiden University) and Professor Spyros Blavoukos (Athens University of Economics and Business): I want to thank them for conveying me their way of thinking and for their insightful comments and suggestions throughout these years.

My research would not have been possible without the contribution of several individuals and institutions. First, I deeply thank all the officials who agreed to be interviewed, they shed light to angles we never knew existed; a special thank you goes to Commissioner Violeta Bulc who instantly agreed to an interview and offered me her own, unique, perspective. Furthermore, the financial support from the Tokyo Foundation for

Policy Research and the Nippon Foundation through the Sasakawa Young Leadership Fellowship Fund (SYLFF), the A. G. Leventis Foundation and the National and Kapodistrian University of Athens through the 'Stavros Tsakyrakis Scholarhsip', was of outmost importance in achieving my research goals.

I would like also to extend my gratitude to Assistant Professors Nassia Yakovaki, Tonia Zervaki, Billy Tsagkroni, and to Dr. Heidi Maurer for their continued support. On a personal note, I would like to thank my family and friends, Dimitris, Fay, Kalliopi, Mary, and Eddie for their constant encouragement and help, which came in different moments and through different means; and to Rodis, for the unwearied and selfless overall support throughout this and other academic journeys, and beyond: a great thank you.

Praise for *EU Climate Diplomacy Towards the IMO and ICAO*

"George Dikaios has addressed in his research a missing link for a better understanding of how climate diplomacy is delivered in the EU. The trinity analytical framework that he is proposing is a fresh and necessary change in the way the EU acts in the international arena. Focus on transport and especially maritime and aviation who are among the biggest polluters, validates his points even stronger. With this book, we got an important tool for better utilization of climate diplomacy and possible ways forward for stronger resilience and climate problem-solving."
—Dr. Violeta Bulc, engineer, innovator, former European Commissioner for Transport

"George Dikaios explores in a comprehensive and fascinating way European climate diplomacy by looking at international transport modes. The book fills a gap in the literature by analyzing contemporary case studies and operationalizing the concept of climate diplomacy. Using extensive primary material including interviews with key officials the author makes a substantial contribution relevant to experts and non-experts alike."
—Professor Spyros Blavoukos, *Athens University of Economics and Business, Greece*

"This excellent book by George Dikaios, based on extensive fieldwork, presents the reader with new and exciting evidence of how the European Union acts in the field of climate diplomacy. It focuses on the aspirations of the EU to influence the international negotiations of utmost importance to contemporary world, over reducing carbon dioxide emissions. By providing a detailed and fascinating account of the EU's efforts to exercise its normative power, this book lets us understand why there has not been any substantial change in this field so far."

—Associate Professor Karolina Pomorska, *Leiden University, The Netherlands*

Contents

Abbreviations and Acronyms

ATAG	Air Transport Action Group
CAEP	Committee on Aviation Environmental Protection
CJEU	Court of Justice of the European Union
CO_2	Carbon dioxide
Commission	European Commission
COP	Conference of the Parties
COREPER	Permanent Representatives Committee
COs	Competence, coordination, coherence
DCS	Data Collection System
DG CLIMA	Directorate General for Climate Action
EAEG	European Aviation and Environment Working Group
EASA	European Union Aviation Safety Agency
EC	European Community
ECAC	European Civil Aviation Conference
EEAS	European External Action Service
EEDI	Energy Efficiency Design Index
ENVI	Committee on the Environment, Public Health and Food Safety
ERTA	European Agreement on Road Transport
EU ETS	European Emissions Trading Scheme
EU	European Union
GHG	Greenhouse gas
ICAO Convention	Convention on the International Civil Aviation Organization
ICAO	International Civil Aviation Organization
IMO Convention	Convention on the International Maritime Organization

IMO	International Maritime Organization
IO	International organization
IPCC	Intergovernmental Panel on Climate Change
LTAG	Long-term Aspirational Goal
MEP	Member of the European Parliament
MEPC	Marine Environmental Protection Committee
MRV	Monitoring, Reporting, Verification
NGOs	Non-governmental organizations
NPE	Normative Power Europe
OECD	Organization for Economic Cooperation and Development
SARPs	Standard and Recommended Practices
SEEMP	Ship Energy Efficiency Management Plan
TEU	Treaty of the European Union
TFEU	Treaty of the Functioning of the European Union
UN	United Nations
UNFCCC	United Nations Framework Convention on Climate Change
USA	United States of America
WA	Work assumption

List of Figures

List of Tables

CHAPTER 1

Introduction and Context

> [On climate change] the EU is like the vegan at the intergovernmental level.[1]

> ICAO and IMO are male organizations; they are not very modern.[2]

Is the European Union (EU) a normative power? Or, moreover, does the EU have any normative power? 2022 marked twenty years from the publication of the seminal article by Ian Manners (2002) titled 'Normative Power Europe: A Contradiction in Terms?', based on which an immense volume of literature has been produced. In its majority, this literature seeks to dredge up the fundamental foundations of the EU's identity, namely whether it is or is not a normative power in its external and foreign affairs. Maybe the reason that Normative Power Europe (NPE) is so appealing—not only in academic literature but also in everyday practice—, apart from the fact that the Union has adopted this characterization for itself, is that the case for being so remains contested and still offers a mosaic of research explorations that open a window *(of opportunity?)* for fruitful discussions about the EU. Acknowledging the above and taking some distance from a systematic analysis of the EU's

[1] Interview no. 49.

[2] Interview no. 40.

© The Author(s), under exclusive license to Springer Nature Switzerland AG 2024

G. Dikaios, *EU Climate Diplomacy Towards the IMO and ICAO*, Palgrave Studies in European Union Politics,
https://doi.org/10.1007/978-3-031-51123-3_1

identity analyses, this book comes to apply the NPE theory as a tool of external action and, in particular, of the diplomatic practice of the EU.

The Union has an ambition to be a normative power in the international system (De Cock, 2011; Huelss, 2012; Pace, 2007). This is evident in the EU Global Strategy published in June 2016, in which it is explicitly stated that the EU should promote and formulate international norms, regarding several issues, to be diffused in the twenty-first century's global governance (European Union, 2016). Federica Mogherini, former Vice President of the European Commission (Commission) and High Representative of the Foreign Affairs and Security Policy of the EU, in 2018 declared that the EU 'will remain the first and most fervent supporter of a rule-based international system, with effective multilateralism at its core' (The Business Time, 2018). By the same token and using more urgent language, her successor and current High Representative Josep Borell has supported that it is vital for the EU's normative power for the former to realize that a competition between different actors for setting global standards is in place. He stressed out the need for the EU to incorporate this development into its policymaking (Borrell, 2021), suggesting the quest for more active and concrete EU efforts in safeguarding its value system, and expressed through its normative power.

One of the topics listed in the EU Global Strategy is that of climate change rules and norms, which are incorporated in an array of policies to be implemented at the global level. Less than a year before the publication of the Strategy, the 2015 Paris Agreement on climate change was signed and the role of the EU was accented as successful (Pomorska & Vanhoonacker, 2016). This gave the EU a new impetus to engage actively with *climate diplomacy*, a term that appears more and more often in most of its official documents and has come to be a prevailing concept to incorporate the action of actors who engage in combating climate change in the political arena through diplomatic means. The tendency towards climate diplomacy was further enhanced by the initiation of the European Green Deal in 2019 by the current President of the Commission, Ursula von der Leyen. One of the defining pillars of the European Green Deal is the leadership role that the EU wishes and needs to perform at a global level to ensure that action in alleviating the challenges posed by climate change will be undertaken by each and every actor that can contribute to this purpose (European Commission, 2019). Consequently, the question that this book seeks to answer is the following: In what ways does the EU perform normative power in the field of climate diplomacy? Through that question and its extensions, the underlining question of whether the EU is or is not a normative power will be also assessed in the end.

To examine whether the EU performs normative power precepts in the area of climate diplomacy, this book will concentrate on the cases of the International Maritime Organization (IMO) and of the International Civil Aviation Organization (ICAO). The reasons behind the choice to analyze these relations are many, but the primary incentive is that nowadays the EU's climate diplomacy is perceived in its broader sense and not only in the context of the United Nations Framework Convention on Climate Change (UNFCCC) nor only in relation to other states (Earsom & Delreux, 2021b; European Parliament, 2018; Minas & Ntousas, 2018). Further, inter-organizational influences are already playing a significant role in combating climate change (Biedenkopf, 2017). Moreover, there is a general acceptance that any climate change measures in both international organizations (IOs) under consideration are a result of the impact that the EU has on them (Earsom & Delreux, 2021a; Lin, 2017; Martinez Romera, 2018).[3] The adoption of the *Carbon Offsetting and Reduction Scheme for International Aviation* (CORSIA) by the ICAO in 2016 and the adoption by the IMO of the *Initial Strategy on Reduction of Greenhouse Gas Emissions from Ships* (Initial Strategy) in 2018 are pivotal in their field and mark a significant change in the way the two IOs approach the fight against climate change. Furthermore, previous Commissioners have stated that the shift in climate policies in both the ICAO and the IMO is a consequence of the EU action (Bulc, 2018; Dimas 2008 in Lindenthal, 2014). On the other hand, EU–IMO and EU–ICAO affairs are generally a rather under-researched subject, especially in regard to climate change, and as such, increasing the focus on their importance can/will offer useful insight into the way the EU performed a normative diplomatic action and influenced the respective IOs to adopt pro-climate strategies. *En passant*, it is worth noting that on the central climate diplomacy webpage of the Commission, the relations with the IMO and the ICAO are absent; they appear in separate specific for international transport webpages,[4] showing

[3] See also the first quote by Commissioner Bulc in page vi; the same opinion was also expressed by many of the interviewees (Interviews no. 19, 22, 25, 27, 28, 31, 32, 38, 41, 42, 44, 45, 47, 54).

[4] For the webpages mentioned, see here: European Commission (2022a; 2022b; 2022c).

on the one hand the importance the Commission gives to their climate action, and on the other hand that these relations seen separate from the overall framework of the EU's climate diplomacy.

Finally, the importance of further research on the topic at hand is highlighted again by the very interesting period that followed after the adoption of the two respective schemes. These schemes called for systematic and serious engagement with the setting of specific targets, the elaboration of future goals, and the review of the initial strategies in order to achieve the expected results and to effectively contribute to the fight against climate change. United Nations (UN) Secretary-General António Guterres, during the Global Sustainable Transport Conference just before the 26th UNFCCC Conference of the Parties that took place in 2021,[5] prompted the two IOs to accelerate their efforts and to put higher targets in emissions cutting, by stating that their current commitments are not expected to deliver adequately to the fight against climate change (The Guardian, 2021). Thus, the period under examination covers the time between the adoption of the Initial Strategy and the CORSIA through to 2021.

1 Research Design and Sources

The research relies primarily on process tracing, which is defined as the attempt to explore 'the chain of events [and] the decision-making process by which initial case conditions are translated into case outcomes' (Van Evera, 1997: 74). This exploration has taken three forms. First, an in-depth analysis of secondary literature has been conducted on a series of topics in order to justify the thematic areas that this book covers: NPE, EU external action, EU climate policy and diplomacy, institutional relations between the EU and the two IOs, climate policy, IMO and ICAO. In particular, scholarly journals and books, articles and contributions to collective books, conference presentations, doctoral theses, as well as newspaper articles that focused on the subject at hand were identified and examined thoroughly and in a comprehensive manner.

Second, a documentary analysis was carried out. Documents relative to this specific policy were retrieved from the official site of the

[5] Conference of the Parties, known by the abbreviation COPs, constitute permanent decision-making institutions of multilateral environmental agreements, responsible for keeping the respective agreements up to date through intergovernmental negotiations.

Commission and other European bodies, as well as from the official sites of the IMO and the ICAO. Specifically, attention was paid to coordination documents, EU treaties, European Parliament documents, Council conclusions, EU–IMO and EU–ICAO press releases and briefings, legislative and official documents retrieved from the IMO and the ICAO, internal reports, and reports from the shipping and the aviation industry. Moreover, the author had the opportunity to consult restricted documents.

Exploring the diplomatic action of the EU regarding climate change towards the IMO and the ICAO in very current times might pose some methodological restrictions to any study. The main challenge is the lack of literature analyzing how the EU approached the two IOs with the aim to put forward its climate agenda. This challenge, though, inevitably leads to the exploration through original sources, which will shed light to the development of the EU's negotiating position and its subsequent externalization. Thus, the third exploratory form used in the research is interviews.

The vast majority of the data that this book relies on originates from fifty-four in-depth qualitative semi-structured interviews with forty-two key officials who participated in the negotiations (either in the EU level, the international level, or both). The interviewees were mainly EU Member States officials (thirty out of forty-two) that traveled to London (IMO Headquarters) or to Montreal (ICAO Headquarters) and were responsible for the negotiation of the EU position to these IOs. Some of the interviewees participated only at the European level negotiations (i.e., at the Working Party on Shipping and the Working Party on Aviation). Additionally, European Commission and European Parliament officers were interviewed, as well as a few EU Member States lobbyists, who again traveled to the respective Headquarters during the negotiations. The tracking of the interviewees was based largely on the official public participating lists of the respective negotiations in the IMO and the ICAO (ICAO, 2013, 2016, 2019; IMO, 2018a, 2018b, 2019, 2021), but the snowball effect was also employed as, especially for the ICAO, locating some key stakeholders was not possible through the official documents, which are confidential.

The interviews were predominantly conducted through phone calls or online meetings, both by reason of the dispersion of the officials interviewed (who were located in 20 countries) and the COVID-19 pandemic, that did not allow for one-to-one meetings but made it easier for the officials to accept online interviews. In addition, an interview of the former European Commissioner of Transport, Violeta Bulc, during whose mandate the Initial IMO Strategy and the CORSIA were adopted, was conducted on the record. The remaining interviews were held off the record for the following reasons. First, the political sensitivity of the issue is *sine qua non* and second, at the time of the interviews and presently, ongoing diplomatic processes are in place where tradeoffs are still on the table. Due to this highly contested arena, the interviewees were reluctant to publicly express their views and, thus, neither their names nor from which country they came from are mentioned in the presentation of the results. It happened also that some of the interviewees asked that direct phrases of them are not to be quoted or to 'take out any insulting comments'.

The above observations lead back to a discussion in the literature concerning the *invisible and unaccountable national representatives* (Juncos & Pomorska, 2011) that are core parts of the decision- and policy-making at the EU level (at the Council of the EU), but for whom we know very little regarding their interaction, positions, and opinions. Thus, by interviewing national representatives who participate both in the EU and the international negotiations, a first image of how they perceive the EU's climate ambition, the EU's internal and external action, as well as how they act in order to fulfill their mandate is drawn. Consequently, a first attempt to make those national representatives visible, and to some extent accountable, is a secondary aim of this book.

The purpose of these interviews was to explore the unique viewpoint of key stakeholders that the primary and secondary sources cannot illustrate adequately; interviews shed light on significantly different aspects of involved actors' motivations and preferences, as well as on their performance and influence strategies within a specific framework (Rathbun, 2008). Moreover, as each interviewee had their own 'truth', the number of interviews allowed for triangulation and the separation of facts from personal views.

The interviews are split into two separate time periods. The ones in the first group were conducted in Spring 2019, while the ones in the second group took place in late 2021–early 2022. The rationale behind doing the first group of interviews arose through interview theory, which argues that 'in instances in which research is being undertaken in a new area without a significant paper trail, a political scientist might consider a set of exploratory interviews to get a better sense of interesting theoretical issues' (Rathbun, 2008: 696). As the institutional background on the EU climate relations with the two IOs under examination, and especially the IMO, was still much under-researched, a series of interviews were completed to sketch the outline of the research field and to allow for an in-depth exploration afterward.[6] As such, references to the interviews are made throughout the book. During the second round, all the interviewees of the first round were contacted again and roughly half of them agreed to be interviewed again.

This book delves specifically into the action of the EU as a whole and it analyzes only the EU-side purposefully, not paying attention, thus, to the inputs of further dynamics that take place within the bargaining arena. Consequently, the role of 'others' or of the interactions of the greater institutional frameworks within which the EU diplomacy unravels is not analyzed. Others, apart from third countries, in the case of the IMO and the ICAO could be also the industry that has a conspicuous position in the respective negotiating and lobbying procedures (Erling, 2017; Interviews no. 26, 43, 48). Furthermore, the book does not focus on the vice versa institutional pressures that might be one explanatory factor of the EU's action. The choice to stick mainly to the EU experience is deliberate and is supported by the following reasons. First, it allows for an in-depth exploration of an up-to-now scarcely researched subject, namely EU climate diplomacy towards third organizations and not the UNFCCC.[7] Second, it digs into the whereabouts of the EU diplomatic apparatus from its foundational pillars to its realization and, finally, to its evaluation. Last but not least, it brings forward an image as it is perceived by the 'protagonists' of the negotiations, who are not career diplomats, but usually officers of the Ministries of Transport (or the respective responsible Agencies) that are expected first to represent

[6] It has to be noted here that IMO officials were also contacted in 2019 but they refused to be interviewed due to the 'neutral' character of the Organization.

[7] While the literature on EU–ICAO relations regarding climate change is voluminous, it has been mostly published in the aftermath of the 2012 inclusion of the aviation emissions in the EU Emissions Trading System (see below) and has mainly analyzed the ramifications of such move and its legal consequences. To the author's knowledge, no articles have systematically analyzed the diplomatic action of the EU towards the ICAO. Even more under-researched is the EU–IMO relations, let alone for their diplomatic relations, with only few exceptions; a very recent exception to the latter is an article by Earsom and Delreux (2021a).

their national interests at the European level and then the European interests at the international level. Before however proceeding with the above, the structure of the book is presented in a short section which follows immediately.

2 Structure of the Book

Following this, first, introductory chapter, the book is organized in the following way. Chapter 2 comes to shape and advance the research field by discussing the existence of the European climate norm, which naturally is bound to be externalized through the NPE approach. It considers NPE within the broader discussion of European external action with a special focus on EU (climate) diplomacy. In this chapter, a systematic reading of the recent literature on EU climate diplomacy takes place, concretely establishing the former as a separate strand of the EU's diplomatic activity. The bond between the internal and the external diplomacy of the EU is also justified.

In Chapter 3, the analytical framework of the book is unfolded and the work assumptions are made. The chapter starts by delineating the performance framework in three-levels: *internal diplomacy*, *external diplomacy*, and *diplomacy evaluation*. First, the internal processes are presented, focusing on the competence conundrum between the EU institutions and the EU Member States; the coordination procedure which takes place within the EU in order to end up in a (common?) negotiating position; and, the coherence (or the single voice) as a result of the two previous steps. Second, the external processes are charted, which contain both the second and the third levels of the analysis, beginning by presenting the factors (or mechanisms) that NPE proposes to explore the EU's external action and continuing to the third level, which reflects on the two previous ones, scrutinizing the coherence of the EU's diplomacy and its effectiveness.

Chapters 4 and 5 consist of the analysis of the case studies of the book. The former unboxes the internal procedures of the EU, exploring the bases on which EU diplomacy is developed. The latter examines the ways through which the EU Member States and institutions externalize the EU climate policy towards the IMO and the ICAO. Both chapters follow a similar structure. Firstly, the information/data gathered through the research period are presented in detail, grouping them under the specific mechanisms that have been selected in the analytical framework. Secondly, the findings are discussed both contextually and methodologically.

Having examined the ways the EU has performed normative power in the field of climate diplomacy in the previous chapters, the book completes with concluding remarks (Chapter 6), in which a broader discussion on the EU's position and normative power in the (climate and IOs) world takes place.

The remainder of this chapter provides the context by outlining the institutional framework(s) within which the research of this book belongs. It draws the picture of how the IMO and the ICAO have engaged in efforts to mitigate greenhouse gas (GHG) emissions from ships and airplanes, respectively, and signposts the role of the EU in these attempts. It then illustrates the position of the EU within these IOs, which presents certain special characteristics, as the EU is not a full member of them. It also further justifies the decision to study the influence of the EU in these organizations through its climate diplomacy, which has been set as a top priority for the current government period in the EU (2019–2024).

3 International Transport Organizations and Climate Change

This section introduces the institutional framework of the IMO and the ICAO, both UN Specialized Agencies, as well as their involvement in global efforts regarding the fight against climate change. It begins by pinpointing the reasons that render both the IMO and ICAO 'unfriendly institutions' (Oberthür, 2006: 59) in the matter of climate change and it continues by showcasing the initial efforts of the EU to make the two IOs engage with the mitigation of GHG emissions and change the 'normality' of their business-as-usual.

Being UN Agencies, both IOs are subject to the global policy guidelines that the UN promotes (White, 1999). Although both Organizations are independent and regulate a specific aspect of the international transport operation, something that does not exactly and directly coincide with the overall work of the UN, they are obliged to follow its overarching political priorities to the extent that the latter find application to their jurisdiction.[8] Thus, when at the UN level mitigation of GHG emissions became a political imperative, it was also diffused to the responsibilities of all IOs linked to the UN which could contribute to achieve this goal.

For both the IMO and the ICAO, the first official declaration of their responsibility to take action in order to tackle climate change and mitigate GHG emissions from the activity that they govern was in the UNFCCC Kyoto Protocol in 1997. The Kyoto Protocol was the first international agreement to pose mandatory obligations to the so-called developed counties for the reduction of greenhouse gas emissions. Although emissions from ships and aircrafts were discussed during the negotiations, an agreement could not be reached due to the incremental growth of the two sectors and the global nature of seafaring and flying (Oberthür, 2003). As both activities take place in locations where no country has direct jurisdiction or responsibilities, the barrier of territoriality and/or nationality, on which the whole UNFCCC system is based on (Corbett & Winebrake, 2010; Erling, 2017), poses several challenges. Therefore, the GHG emissions reductions by shipping and aviation activity were delegated to the respective organizations through Article 2(2) of the Kyoto Protocol: 'The Parties included in Annex I shall pursue limitation or reduction of emissions of greenhouse gases not controlled by the Montreal Protocol from aviation and marine bunker fuels, working through the International Civil Aviation Organization and the International Maritime Organization, respectively'.

International Maritime Organization and Climate Change

The *Intergovernmental Maritime Consultative Organization,* the percussor of the IMO, was established in 1958 according to the *Convention of the Intergovernmental Maritime Consultative Organization* which was signed in 1948. Its name was changed to *International Maritime Organization* in 1982. The fundamental objective of the IMO is to institutionalize the cooperation among its Member States regarding regulations and all the relevant technical practices related to international trade shipping, with the aim to eliminate any disclinations or restrictions

[8] For the ICAO, see Sochor (1991: 39–40). For the IMO, Article 1(d) of the IMO Convention: The purposes of the Organization are: (d) to provide for the consideration by the Organization of any matters concerning shipping that may be referred to it by any organ or Specialized Agency of the United Nations.

that may pose barriers to the latter. Special attention to the Organization's purposes is given to 'maritime safety, efficiency of navigation and the prevention and control of marine pollution from ships' (Article 1(a) of the IMO Convention). The IMO has 175 Member States and three associated members (i.e., Faroes, Hong Kong, and Macao).

The institutional structure of the IMO consists of the Assembly, the Council, the Secretariat, five main Committees, and many more sub-committees. All Member States of the IMO participate in the Assembly, with equal voting rights; it convenes every two years and is responsible for the long-term planning of the Organization, its budget and financing, as well as electing the members of the Council. The Council covers for the Assembly when it is not in session and its mandate covers almost all the responsibilities of the latter. It consists of 40 members, which are split into three categories: '10 states with the largest interest in providing shipping services; 10 states with the largest interest in international seaborn trade; and 20 states that ensure the representation of all major geographic areas of the world'. One of the main tasks of the Council is to 'receive reports and proposals of the Committees and submit them to the Assembly and Member States'. What is interesting about the Committees of the IMO is that their membership consists of all the Member States of the Organization, a characteristic that also applies to the sub-committees that are open again to all Member States (IMO, 2022c).

The Committee relevant to the field of this book is the Marine Environment Protection Committee (MEPC), which is mandated to coin rules and adequate amendments to conventions regarding the pollution caused by the shipping activity.[9] The MEPC is, therefore, charged with finding the necessary consensus and means to achieve it for the currently burning topic of tackling climate change through international shipping. The story of this mandate starts in the mid-1990s, around the same period that the Kyoto Protocol was adopted, followed by an MEPC request to the IMO Secretariat to prepare a comprehensive study on GHG emissions from ships. Since 1999, when the *First IMO Study on Greenhouse Gas* was released, the IMO regularly conducts related research. In 2020, the *Fourth IMO Greenhouse Gas Study* was published, indicating that, for 2017, the contribution of international shipping in

[9] It is noted that the MEPC presides over subjects that surpass its original purpose (Article 38 of the IMO Convention), seen it is name, as it also regulates air pollution. This is an oxymoron as the marine and air environments are very different in nature.

GHG emissions globally was 2.97 per cent and for 2018 was 2.89 per cent; recording an increase of about 10 per cent in emission tonnes in seven years, since 2012 (IMO, 2020). This is also the largest percentage since 1996,[10] while all the projecting scenarios accumulated in the *Study* show an upward tendency from 2018 to 2050 (IMO, 2020: 236).

The attempts at the diplomatic level to find solutions on the issue of GHG emissions expand throughout these—almost—three decades. In the *First IMO Study*, a package of different measures was put forward that until today dominates the negotiations. Resolution A.963(23) was adopted in 2003 pointing towards the evolution of a system for the monitoring and reporting of the GHG emissions, based on technical and operational measures, as well as on market-based measures (IMO, 2003; cf. Agarwal, 2009). From 2003 to 2011,[11] when the first germinal measures regarding climate change were adopted by the IMO (Harrison, 2017; Shi & Gullet, 2018), a work plan programming the development of the respective measures, as well as guiding principles for the future action of the Organization to regulate the reduction of the GHG emissions were adopted (Zhang, 2016). The principles adopted in 2008, which try to concurrently conciliate climate measures for all, the freedom of international trade, and the just distribution of the implementation weight among IMO's Member States, institutionalized an underlying conflict on the ways a reduction scheme should be developed and led the negotiations for the following decade (Doelle & Chircop, 2019).

The measures adopted in 2011 focused only on the technical and operational measures, as an agreement on the market-based measures could not be reached. To solve the gridlock, the MEPC created the Intersessional Meeting of the Working Group on Reduction of GHG Emissions from Ships in order to facilitate the negotiation process, to work systematically on the subject, and to make suggestions for an effective mechanism

[10] The First IMO Greenhouse Gas Study was published in 1999. It estimated the contribution of the shipping industry to gas emission at 1.8 per cent in 1996; the Second in 2009 at 2.7 per cent for 2007 (later corrected to 2.8 per cent); and the Third in 2014 at 2.2 per cent for 2012, recognizing the impact of the global financial crisis (IMO, 2015). The average percentage of the CO_2 emissions between 2007 and 2012 was 2.6 per cent (IMO, 2017), while the average between 2012 and 2018 was 2.83 per cent (IMO, 2020).

[11] For an analysis on the development of the institutional governing architecture of the GHG emissions in IMO up until 2010, see Hackmann (2012).

that would contribute to the mitigation of GHG emissions (Zhang, 2016).

Nevertheless, the regulatory measures inserted into the IMO system in 2011 were the first mandatory and essential (yet, very small) steps towards controlling GHG emissions from ships (IMO, 2022a; Shi & Gullet, 2018). The two mechanisms that were developed, the Energy Efficiency Design Index (EEDI) and the Ship Energy Efficiency Management Plan (SEEMP), are still in force until today. The EEDI concerns the building of specific types and the size of new ships under specific requirements following the latest developments of technology that would allow for energy efficient tools with an ultimate goal of less emissions. The SEEMP applies to all the existing and new ships by concentrating on the ways they operate and on how, through possible changes in their operation, and their energy performance could be improved. The SEEMP is voluntary and does not determine targets, meaning that its implementation does not affect the operation of the current global fleet. In contrast, whereas the EEDI produces legally binding responsibilities, it allows, at the same time, shipbuilders and ship designers to develop technologies that they believe will serve the purpose of reducing GHG emissions. Moreover, although in the IMO the decision-making usually takes place with consensus, for the EEDI a vote was required, with several developing countries downvoting the respective regulation, undermining it from the beginning of its implementation.

The aftermath of the 2015 Paris Climate Agreement created favorable circumstances (Corbett et al., 2020; Earsom & Delreux, 2021a) and reintroduced conference diplomacy back to a positive framework (Brun, 2016). In this context, and with voices though calling for the full decarbonization of the shipping sector (Traut et al., 2018), the IMO made an ostensibly big step. In 2018, it adopted the *Initial Strategy on Reduction of Greenhouse Gas Emissions from Ships*, which set the target of mitigation at 50 per cent by 2050, by the international shipping activity. Once more, a voting procedure took place with only 100 out of 175 IMO Member States voting in favor of the Strategy, raising coherence issues among the Member States, as well as uncertainties about the future realization of the Strategy and its contribution to the fight against climate change. Concerns grew taking into account the nature of the Initial

Strategy, which is just a political declaration without legally binding consequences,[12] in addition to scientific evidence that full decarbonization of the shipping sector would be able to be achieved by 2035 (International Transport Forum, 2018).

International Civil Aviation Organization and Climate Change

ICAO was established in 1947 based on the *Chicago Convention on International Civil Aviation* that was signed in 1944. The ICAO's main goals are the development of the principles and techniques of international aviation and the design and development of aviation transportation and flight safety measures, the promotion of the use of innovative technology, and the development of all aspects of international civil aviation (Article 44 of the ICAO Convention).[13] Today, it has 193 Member States, and its main task is to promote the diplomatic cooperation in every aspect that touches upon civil aviation globally. According to its own description of its function, the ICAO develops and sets in motion standards that guide the air transport operation internationally. These standards, a result of consensus between the Member States of the Organization, do not transcend national regulations but, rather, provide the framework within which sovereign states ought to work, in order to offer access to a 'safe, secure and sustainable' environment for civil aviation (ICAO, 2022a).

Its institutional arrangement consists of two diplomatic bodies, the ICAO Assembly composed of all the Member States of the Organization and the ICAO Council where 36 Members participate, as well as the ICAO Secretariat, which is responsible for the everyday functioning of the Organization. As far as it concerns the main governing organs of the

[12] For an evaluation of the Initial IMO Strategy, see Doelle and Chircop (2019).

[13] Objectives: The aims and objectives of the Organization are to develop the principles and techniques of international air navigation and to foster the planning and development of international air transport so as to: (a) Insure the safe and orderly growth of international civil aviation throughout the world; (b) Encourage the arts of aircraft design and operation for peaceful purposes; (c) Encourage the development of airways, airports, and air navigation facilities for international civil aviation; (d) Meet the needs of the peoples of the world for safe, regular, efficient, and economical air transport; (e) Prevent economic waste caused by unreasonable competition; (f) Insure that the rights of contracting States are fully respected and that every contracting State has a fair opportunity to operate international airlines; (g) Avoid discrimination between contracting States; (h) Promote safety of flight in international air navigation; (i) Promote generally the development of all aspects of international civil aeronautics.

ICAO, in keeping with the majority of the IOs, the Assembly acts as the main deliberative body, where the general guidelines of its function and the evolution of air transport regulations are decided through the system *one state–one vote*. It is responsible, in rough lines, for the overall overview of the Organization's work and the election of the Member States to participate in the ICAO Council. The members of the Council represent at the same time those States which have a great interest in international aviation and the UN regional groups,[14] so there is fair distribution around the world (ICAO, 2022c). As the Assembly meets every three years (unless extraordinary meetings are requested), it decides upon the subjects that the Council and the Secretariat will work on for the coming three-year period and, during the next session, it approves, amends, or rejects their proposals on all the technical, political, socio-economic, and any other issues that have been prepared throughout this period.[15]

Therefore, the ICAO Council has a three-year mandate to negotiate international policies to be adopted by the Assembly. Its significance for the ICAO and civil aviation in general is crucial as it specifies everything relevant to the international flights and its regulatory background through the so-called Standard and Recommended Practices (SARPs). Additionally, in 2014, the Council put forward the *No Country Left Behind* initiative that aims to assist all the ICAO Member States to implement the totality of ICAO's policies in order for consistent rules applying to aviation internationally (ICAO, 2022b).

To assist its broad scope of work, the Council has established several committees that report to it and prepare—again, during the three-year period—the relevant to their role SARPs. One of these technical committees is the Committee on Aviation Environmental Protection (CAEP), which was established in 1983 to devise rules concerning mainly aircraft noise and emissions.[16] As a result, since the 1990s, when the reduction of

[14] Representatives in UN bodies usually originate from the five UN Regional Groups (namely, African States, Asia–Pacific States, Eastern European States, Latin American and Caribbean States, and Western European and other States) in order to provide geographical representation (Aeschlimann & Regan, 2017).

[15] For in-depth analyses on the institutional framework of the ICAO, see cf. Weber (2021).

[16] Again, air pollution by aircrafts has been somehow arbitrarily added to the general 'environmental problems', much in accordance with the case of emissions in the IMO. Climate change is, once again, viewed as an environmental problem, rather than as a separate and more rounded one.

GHG emissions became a political priority for world politics, CAEP has been delegated to research and devise recommendations on how international aviation will contribute its fair share to the global mitigation efforts.

Mitigation of GHG emissions in aviation was institutionally established in the UNFCCC Kyoto Protocol in 1997, where it was stated that the ICAO ought to take action towards regulating aviation emissions. It was, of course, before that that scientific evidence was pointing towards the emissions from aircrafts being a problem for the atmosphere, characterizing GHG emissions as the *emerging issue* of the decades to come.[17] Since then, a slew of scientific documents has been published, highlighting the negative effects of aircraft emissions on climate (cf. Amanatidis & Friedl, 2004; Brasseur et al., 1998), with flagship publication of the *Aviation and the Global Atmosphere* in 1999, a special report prepared by the Intergovernmental Panel on Climate Change (IPCC) in collaboration with the Scientific Assessment Panel to the Montreal Protocol on Substances that Deplete the Ozone Layer (Penner et al., 1999). The report was commissioned by the ICAO and it put the share the aviation sector has on global warming on the table. Nevertheless, it was only in 2010 that the ICAO adopted preliminary decisions regarding fuel efficiency, which were reaffirmed in the next Session of the Assembly in 2013 (Weber, 2021).

In the 37th Session of the Assembly in 2010, analysts spotted the foundations of the global strategy that was adopted in the 39th Session of the Assembly in 2016, titled CORSIA (Abeyratne, 2017; Weber, 2021). The preparation of the scheme was allocated by the Council to the CAEP after the 38th Session of the Assembly in 2013, which indicated the need for a more comprehensive carbon policy of the Organization (Erling, 2017; Weber, 2021). The selection of an offsetting scheme which allows trading emissions outside a particular sector prevailed over an emissions trading system within the sector, which was included in the initial proposals but was removed, since the underlying rationale of the former is that the

[17] Cf. the collective report titled *Global Atmospheric Effects of Aviation: Report of the Proccedings of the Symposium* (Virgina Beach, Virginia USA, 15–19 April 1996), sponsored by Association of European Research Establishments in Aeronautics; European Commission; Intergovernmental Panel on Climate Change; ICAO; National Aeronautics and Space Administration, USA; National Oceanic and Atmospheric Administration, USA; United Nations Environment Programme; and World Meteorological Organization.

emissions can grow uncontrollably and can be counterbalanced by other activities in different sectors. Therefore, the offsetting scheme is a system that serves the aviation interests, and it does not contribute at all to the reduction of GHG emissions (Abeyratne, 2017; Erling, 2017; Lyle, 2018).

The implementation schedule of the CORSIA is split into three periods: from 2021 through 2023, from 2024 through 2026, and from 2027 through 2035. The introductory period is called *pilot phase* and is addressed to States-volunteers that wish to develop an offsetting system in their territory. Next comes the *first phase*, in which the States-volunteers of the pilot phase participate, as well as additional States that wish to volunteer. In this phase the development of the offsetting scheme begins. Last, the *second phase* includes all States that have a certain aviation activity (with exceptions to several categories of the developing countries unless they volunteer).[18]

In its Session of the Assembly in 2019, the ICAO did not go any further with the CORSIA, pending the start of its enactment period in 2021. Instead, it requested the ICAO Council to continue working on and looking into the specifics of its implementation (Abeyratne, 2020). Concurrently, the Organization distributes related SARPs and guidelines for the best possible preparations and implementation practices (Weber, 2021).

During the 2019–2022 term, the most heated discussion in CAEP was about the Long-Term Aspirational Goal (LTAG) for reduction of GHG emissions, with a couple of negotiators claiming that from the moment the IMO managed to set one, the ICAO needs to do the same (Interviews no. 28, 34).[19] The importance of this activity becomes evident as, according to the latest scientific evidence, aviation contribution to the global emissions varies between 2.1 and 2.5 per cent; some even argue for triple percentages if other GHGs are part of the equation (ATAG, 2020; Lee, 2017; Ritchie, 2020). For the EU, 'if global aviation were a

[18] ICAO (2016); see also Lee (2017); for analyses on how the CORSIA works, see Erling (2017); Weber (2021). As of July 2023, 115 States participate voluntarily in CORSIA, representing more than 90 per cent of global aviation.

[19] The LTAG was adopted in the 2022 ICAO Assembly, but again it did not set any 'specific obligations or commitments in the form of emissions reduction goals to individual States' (ICAO, 2023).

country, it would rank in the top 10 emitters' (European Commission, 2022c).

The EU in the IOs

The need for urgent and more ambitious action from both the IMO and the ICAO to achieve the global goal of upholding the rise of the temperature is highlighted by the literature (Bullock et al., 2022; Lyle, 2018). Within this context, this book explores the role of one of the most publicly recognized ambitious and exceptionally active actors in the field, namely the EU.[20] According to the officials interviewed, it is dreadful to realize that the ICAO and the IMO believe that they are excluded from the Paris Agreement because they are not mentioned explicitly within the official text (in contrast to the Kyoto Protocol) (IMO, 2022b); for the EU, both Organizations need to put similar efforts in accordance to the rest of the world, regardless of formalities (Interviews no. 28, 35, 40, 52).

The EU has long been engaged with developing solutions to fight climate change from the shipping and aviation activity, especially after the Kyoto Protocol was adopted and even more so around the period that it entered into force in 2008. While the Union's major and better-known efforts in the mid-2000s concerned the aviation sphere with the unilateral act of including international flights in the EU Emissions Trading System, the EU had also—from early on—showed its intentions to undertake substantial action towards international shipping as well (Corbett & Winebrake, 2010). The reason for the development of the related systems was the 'dissatisfaction' with the ICAO and the IMO (Lindenthal, 2014; Van Leeuwen, 2015; Van Leeuwen & Kern, 2013), which are often judged as slow and with dubious results in the GHG emission action (Martinez Romera & van Asselt, 2015; Nengye & Maes, 2012; Oberthür, 2006; Shi, 2018; Interviews no. 14, 15, 18).

Moreover, the EU has developed one of the most comprehensive systems around the world to mitigate GHG emissions, namely the European Emissions Trading System (EU ETS). Simply put, the EU ETS regulates the GHG emissions of a broad array of industrial and power sectors in the EU. The Union sets the maximum levels of allowances for

[20] In the literature on IMO, the instrumental role of the Marshall Islands is pointed out as well (Corbett et al., 2020; Earsom and Delreux, 2021a), while in the literature of ICAO, no other actor seems to prevail rather than the EU.

companies belonging to the related sectors. In case a company exceeds the stipulated margin, it is obliged to buy further allowances to cover its surpass. Buying can be done either from the open market or by other companies that have emitted less than their ordered level. The goal of the scheme is to provide financial incentives to achieve effective GHG emission reduction at the lowest possible cost (cf. Ellerman, 2009; Watanabe & Robinson, 2005). Operating the scheme with a certain level of success (Vis 2006 at Skjærseth & Wettestad, 2008b: 276), the EU decided to unilaterally include international flights, along—of course—with all the intra-European ones, in the EU ETS. For that purpose, it adopted the Directive 2008/101/EC (Official Journal of the European Union, 2009), which was to be effective from 2012 and would regulate GHG emissions from all flights arriving or departing from any European airport. This decision was faced with fierce resistance by third countries, which also brought up the subject to the ICAO and activated legal procedures in front of the Court of Justice of the European Union (CJEU) to challenge the Directive as going against international law.[21] Although the CJEU decided in favor of the Directive, the EU succumbed and froze its implementation by introducing three consecutive decisions known by the characteristic title 'Stop the Clock'; the first stopped the implementation of the Directive for one year, the second for the period between 2014 and 2016 and the third from 2017 to 2023. The reason behind this decision was the political outcry that the EU as a whole experienced within the ICAO. The Commission framed its decision as a step towards giving space to the ICAO to negotiate without ultimatums (European Commission, 2012). One of the interviewees, who participated at the time to one of the European delegations that traveled to the ICAO, recounted an image of 'unbelievable diplomatic defeat' (Interview no. 53). Since then, the relations between the involved parties have smoothened (or normalized) and multilateral diplomacy has been reinstated as a means for cooperation.

[21] Directive 2008/101/EC resulted to a spate of academic literature examining mostly its legal ramifications (De Paula Domingos, 2012; Dejong, 2013; Fahey, 2012; Hertogen, 2012; Martinez Romera, 2018; Petersen, 2008; Reagan, 2008), evaluating its general consequences and its potential implementation outcomes (Malina et al., 2012; Preston et al., 2012; Prum & Kisska-Schulze, 2015; Scott & Rajamani, 2012). Interestingly enough, the majority of the papers cited here record a connection between the inclusion of the international aviation in the EU ETS with the ICAO's reaction to develop CORSIA.

Unlike to what was the case with the ICAO, the EU followed a milder approach towards the IMO, which coincided in more ways with what the IO was doing at the time. In other words, the EU adapted more to the reality of the IMO. When the Organization adopted the EEDI, the EU pursued it and developed its own monitoring, reporting, and verification (MRV) scheme, established under the Regulation (EU) 2015/757, known under the acronym MRV Regulation (Official Journal of the European Union, 2015). Pursuant to this Regulation, MRV of CO_2 emissions from maritime transport includes all large ships using EU ports and consists of the first binding stage of recording the emissions track from the shipping activity in order to understand the level of fuel consumption that takes place within the EU. The Regulation is in effect since 1 January 2018. It has to be mentioned that, in parallel, the IMO was developing a corresponding system with a similar goal, which came into force one year later, on 1 January 2019, under the name Data Collection System (DCS). The DCS applies to all ships under SEEMP. A significant difference between the two systems relates to their *transparency* feature; in the DCS the data are publicly available, while in the SEEMP they are bound by confidentiality (cf. Rony et al., 2019).[22] On the eve of the adoption of the Initial IMO Strategy, the EU also adopted Directive 2018/410 pointing out the firm EU commitment in reducing GHG emissions from the shipping industry, as well as urging the IMO to be particularly ambitious towards this purpose. Although, according to the EU, the IMO performed limited progress throughout the years and with significant delays, its approach remained multilateral with an aim of finding a collective solution to the issue at stake.

The different levels of intensity in the approach of the EU towards the ICAO and the IMO draw parallels to the efforts the EU has put to advance its position within the two IOs. At this point, it is essential to mention that the EU holds only an observer status in both IOs. However, there are some deviations between the two systems. The Commission of the—back then—European Community[23] became an observer in the ICAO in 1989, a position which offered the ability to the Commission

[22] Other authors argue that neither of the systems encourage transparency, but that the EU MRV system is in a better position in this regard, compared to the IMO DCS (Deane et al., 2019).

[23] For purposes of clarity in continuity, the term 'EU' is applied for all the periods of the European Community.

to be present on an ad hoc basis in some of the meetings that took place within the ICAO. It has to be noted here that air transport in the EU is governed under the status of the shared competences. Therefore, as air transport policies boomed within the EU (Kassim & Stevens, 2010), the Commission realized that its voice was not heard at all at the international fora where all the decisions on the international regulations were adopted; the latter significantly affected the EU's internal policy. Following the publication in 2001 of the White Paper titled *European Transport Policy for 2010: Time to Decide* (Commission of the European Communities, 2001), which raised the above issue and intertwined it with the impotent observer status in the IOs responsible for the international transport regime, the Commission began to push for an enhancement of its position. It is interesting to note that the White Paper, apart from arguing in favor of a single EU voice, also put forward the promotion of sustainable development to the transport sector, via the activity of the EU within the respective IOs. Through pressuring, while considering that the Chicago Convention allows only for states to become contracting parties to the Convention, the Commission managed to inaugurate a permanent office in the ICAO Headquarters in 2005, maintaining, though, its observer status. Still, the EU's presence and influence on the ICAO has grown in the course of the years past, through the former's active involvement in the shaping and implementation of an array of technical matters concerning aviation, as well as through its involvement in promoting environmental requirements (Vaugeois, 2019).

In the case of the IMO, notwithstanding the Commission being an observer of the organization since 1974 (15 years prior to engaging with the ICAO), the relations between the two bodies have been more neutral over the years. This 'delay' could be attributed to the elevation of interest in maritime affairs on behalf of the Union which only happened during the 1990s (Pallis, 2002), when the Commission gained control over subjects deliberated at the IMO. After the release of the White Paper, the Commission promoted a request to the IMO regarding its status, without making progress. Thus, the EU Representative residing in the IMO has for long acquired an *invisible role* which has been limited to the coordination of the European position among the EU Member States and to the monitoring of their action (Wouters et al., 2009).

Because of the unique hybrid presence of the Commission in the IMO and the ICAO, challenges for the EU Member States' actions arise, with the Commission resorting to the CJEU to crystallzse the limits of unilateral activity conducted by the Member States. In the case of C-45/07, the CJEU adjudicated in favor of the Commission by securing that, in the domains that exclusive competences have been transferred to the EU level, Member States cannot act on their own in any context and framework, whether this is national, European or international.[24] In addition, according to the EU Treaties, the Member States may only differentiate their position in the case that the *acquis* is not jeopardized (Van der Mei, 2016; Wouters et al., 2009).

It is interesting to note that several interviewees raised the issue of clarifying/justifying of whom we mean when we use the term 'EU': the EU institutions, the EU Member States, or the EU as a whole. This stance further adds to the conflict between the institutions and the Member States, in spite of their concurrent perception of European success in both Organizations, concerning climate change. To that end, within the corpus of the research at hand, distinct references are made wherever needed and the term 'EU as a whole' is applied widely.[25]

Against the backdrop sketched in this introductory chapter, the book examines both the internal and external diplomacy that is performed within and by the EU as a whole towards the IMO and the ICAO. The analysis of the EU's diplomatic performance is entwined with the NPE approach, allowing for an exploration of how the EU acts in order to alter the 'normal' of its policies' receiving end.

List of Interviews

Interview no. 14, Advocate of the industry of an EU Member State, phone call, 7 May 2019.
Interview no. 15, European Parliament official, phone call, 7 May 2019.
Interview no. 18, European Commission official, phone call, 22 May 2019.
Interview no. 19, European Commission official, phone call, 22 May 2019.
Interview no. 22, EU Member State official, phone call, 24 May 2019.
Interview no. 25, EU Member State official, virtual call, 2 November 2021.

[24] Case-45/07 regarded a dispute between the Commission and the Hellenic Republic on the matter of maritime safety and the failure of the Member State to fulfill its obligations under the European *acquis*, which jeopardized the division of powers between the two actors (Court of Justice of the European Union, 2009).

[25] 'EU as a whole' was also applied during the recent announcements of the European Commission relating to the ongoing Ukrainian Crisis, a rather uncommon occurrence.

Interview no. 26, Advocate of the industry of an EU Member State, virtual call, 10 November 2021.
Interview no. 27, EU Member State official, virtual call, 11 November 2021.
Interview no. 28, EU Member State official, virtual call, 16 November 2021.
Interview no. 31, EU Member State official, virtual call, 17 November 2021.
Interview no. 32, EU Member State official, virtual call, 18 November 2021.
Interview no. 34, EU Member State official, virtual call, 26 November 2021.
Interview no. 35, European Parliament official, virtual call, 26 November 2021.
Interview no. 38, EU Member State official, virtual call, 1 December 2021.
Interview no. 40, European Commission official, virtual call, 3 December 2021.
Interview no. 41, EU Member State official, virtual call, 3 December 2021.
Interview no. 42, EU Member State official, virtual call, 3 December 2021.
Interview no. 43, EU Member State official, virtual call, 6 December 2021.
Interview no. 44, EU Member State official, virtual call, 7 December 2021.
Interview no. 45, EU Member State official, virtual call, 7 December 2021.
Interview no. 47, European Commission official, virtual call, 8 December 2021.
Interview no. 48, Violeta Bulc, European Commissioner of Transport (2014–19), virtual call, 20 December 2021.
Interview no. 49, Advocate of the industry of an EU Member State, phone call, 3 January 2022.
Interview no. 52, European Commission official, virtual call, 24 January 2022.
Interview no. 53, European Parliament official, in person, 31 January 2022.
Interview no. 54, European Commission official, virtual call, 10 February 2022.

References

Abeyratne, R. (2020). The Outcome of the 40th ICAO Assembly: A New Look at ICAO? *Air & Space Law, 45*(1), 81–96.

Abeyratne, R. (2017). Outcome of the 39th Session of the International Civil Aviation Organization Assembly. *Air & Space Law, 42*(1), 13–28.

Aeschlimann, J., & Regan, M. (2017). *The GA Handbook: A Practical Guide to the United Nations General Assembly* (2nd ed.). Permanent Mission of Switzerland to the United Nations.

Agarwal, S. K. (2009). Mitigating Greenhous Gas (GHG) Emissions from International Shipping in Post-Kyoto Climate Policy: Legal Issues and Challenges. *Maritime Affairs: Journal of the National Maritime Foundation of India, 5*(1), 73–96.

Amanatidis, G., & Friedl, R. (2004). Science Related to Atmospheric Effects of Aircraft Emissions Continue to Mature. *ICAO Journal, 59*(5), 14–15.

ATAG. (2020). *Aviation Benefits Beyond Borders: Powering Global Economic Growth, Employment, Trade Links, Tourism and Support for Sustainable Development through Air Transport, Despite Global Crisis*. Air Transport Action Group.

Biedenkopf, K. (2017). Relations between International Organizations in Combating Climate Change. In R. Biermann & J. A. Koops (Eds.), *The Palgrave Handbook of Inter-organizational Relations in World Politics* (pp. 649–676). London: Palgrave Macmillan.

Borrell, J. (2021, November 12). A Strategic Compass for Europe. *Project Syndicate*. https://www.project-syndicate.org/commentary/eu-strategic-compass-by-josep-borrell-2021-11. Last accessed: 22 March 2022.

Brasseur, G., Amanatidis, G. T., & Angeletti, G. (Guest Eds.). (1998). European Scientific Assessment of the Atmospheric Effects of Aircraft Emissions. *Atmospheric Environoment, 32*(13), 2327–2422.

Brun, A. (2016). Conference Diplomacy: The Making of the Paris Agreement. *Politics and Governance, 4*(3), 115–123.

Bulc, V. (2018, July 12). Shipping Is Delivering on Climate Change. *The Parliament Magazine*. https://www.theparliamentmagazine.eu/news/article/shipping-is-delivering-on-climate-change. Last accessed: 13 March 2022.

Bullock, S., Mason, J., & Larkin, A. (2022). The Urgent Case for Stronger Climate Targets for International Shipping. *Climate Policy, 22*(3), 301–309.

Commission of the European Communities. (2001). *White Paper—European Transport Policy for 2010: Time to Decide*, COM(2001) 370 final.

Corbett, J. J., & Winebrake, J. J. (2010). The Role of International Policy in Mitigating Global Shipping Emissions. *The Brown Journal of World Affairs, 16*(2), 143–158.

Corbett, J., Ruwet, M., Xu, Y. C., & Weller, P. (2020). Climate Governance, Policy Entrepreneurs and Small States: Explaining Policy Change at the International Maritime Organization. *Environmental Politics, 29*(5), 825–844.

Court of Justice of the European Union. (2009). *Judgement of the Court (Second Chamber) of 12 February 2009. Commission of the European Communities v Hellenic Republic. Failure of a Member State to fulfil obligations—Article 10 EC, 71 EC and 80(2) EC—Maritime safety—Monitoring of ships and port facilities—International agreements—Division of powers between the Community and the Member States. Case C-45/07*. https://curia.europa.eu/juris/liste.jsf?language=en&num=C-45/07. Last accessed: 30 November 2022.

De Cock, G. (2011). The European Union as a Bilateral 'Norm Leader' on Climate Change vis-à-vis China. *European Foreign Affairs Review, 16*(1), 89–105.

De Paula Domingos, P. (2012). Fighting Climate Change in the Air: Lessons from the EU Directive on Global Aviation. *Revista Brasileira de Política Internacional, 55*(special edition), 70–87.

Deane, F., Huggins, A., & Karim, S. M. (2019). Measuring, Monitoring, Reporting and Verification of Shipping Emissions: Evaluating Transparency and Answerability. *Review of European, Comparative & International Environmental Law, 28*(3), 258–267.

Dejong, S. M. (2013). Hot Air and Hot Heads: An Examination of the Legal Arguments Surrounding the Extension of the European Union's Emissions Trading Scheme to Aviation. *Asian Journal of International Law, 3*(1), 163–188.

Doelle, M., & Chircop, A. (2019). Decarbonizing International Shipping: An Appraisal of the IMO's Initial Strategy. *Review of European, Comparative & International Environmental Law, 28*(3), 268–277.

Earsom, J., & Delreux, T. (2021a). A Nice Tailwind: The EU's Goal Achievement at the IMO Initial Strategy. *Politics and Governance, 9*(3), 401–411.

Earsom, J., & Delreux, T. (2021b). Evaluating EU Responsiveness to the Evolution of the International Regime Complex on Climate Change. *International Environmental Agreements: Politics, Law and Economics, 21*(4), 711–728.

Ellerman, D. (2009). The EU's Emissions Trading Scheme: A Prototype Global System? *MIT Joint Program on the Science and Policy of Global Change*, Report no. 170.

Erling, U. M. (2017). International Aviation Emissions Under International Civil Aviation Organization's Global Market Based Measure: Ready for Offsetting? *Air & Space Law, 42*(1), 1–12.

European Commission. (2012). *Stopping the Clock of ETS and Aviation Emissions Following Last Week's International Civil Aviation Organization (ICAO) Council*. MEMO/12/854.

European Commission. (2019). *Communication from the Commission to the European Parliament, the European Council, the Council, the European Economic and Social Committee and the Committee of the Regions: The European Green Deal*, COM(2019) 640 final.

European Commission. (2022a). *International Action on Climate Change*. https://ec.europa.eu/clima/eu-action/international-action-climate-change_en. Last accessed: 28 May 2022.

European Commission. (2022b). *Reducing Emissions from Aviation*. https://ec.europa.eu/clima/eu-action/transport-emissions/reducing-emissions-aviation_en. Last accessed: 28 May 2022.

European Commission. (2022c). *Reducing Emissions from the Shipping Sector*. https://ec.europa.eu/clima/eu-action/transport-emissions/reducing-emissions-shipping-sector_en. Last accessed: 28 May 2022.

European Parliament. (2018). *Resolution of 3 July 2018 on Climate Diplomacy*. 2017/2272(INI).

European Union. (2016, June). *Shared Vision, Common Action: A Stronger Europe—A Global Strategy for the European Union's Foreign and*

Security Policy. http://www.eeas.europa.eu/archives/docs/top_stories/pdf/eugs_review_web.pdf. Last accessed: 13 March 2022.

Fahey, E. (2012). The EU Emissions Trading Scheme and the Court of Justice: The 'High Politics' of Indirectly Promoting Global Standards. *German Law Review, 13*(11), 1247–1268.

Hackmann, B. (2012). Analysis of the Governance Architecture to Regulate GHG Emissions from International Shipping. *International Environmental Agreements: Politics, Law and Economics, 12*(1), 85–103.

Harrison, J. (2017). *Saving the Ocean Through Law: The International Legal Framework for the Protection of the Marine Environment*. Oxford University Press.

Hertogen, A. (2012). Sovereignty as Decisional Independence over Domestic Affairs: The Dispute over Aviation in the EU Emissions Trading System. *Transnational Environmental Law, 1*(2), 281–301.

Huelss, H. (2012). A Force for Good Governance? The European Union's Normative Power and Standards of Appropriate Governing. *European Foreign Affairs Review, 17*(1), 93–112.

ICAO. (2013). *38th Session of the ICAO Assembly, 24 September to 4 October 2013, Montréal: List of Delegates No. 10.0.*

ICAO. (2016). *39th Session of the ICAO Assembly, 27 September to 7 October 2016, Montréal: List of Delegates as of 06-Oct-2016 16:33 hrs.*

ICAO. (2019). *40th Session of the ICAO Assembly, 24 September to 4 October 2019, Montréal: List of Delegates as of 11-Oct-2019 16:16 hrs.*

ICAO. (2022a). *About ICAO*. https://www.icao.int/about-icao/Pages/default.aspx. Last accessed: 25 January 2022.

ICAO. (2022b). *No Country Left Behind*. https://www.icao.int/about-icao/nclb/pages/default.aspx. Last accessed: 26 January 2022.

ICAO. (2022c). *The ICAO Council*. https://www.icao.int/about-icao/Council/Pages/Council.aspx. Last accessed: 26 January 2022.

ICAO. (2023). *Long Term Global Aspirational Goal (LTAG) for International Aviation*. https://www.icao.int/environmental-protection/Pages/LTAG.aspx. Last accessed: 8 August 2023.

IMO. (2003, March 4). *Resolution A.963(23) Adopted on 5 December 2003 (Agenda Item 19)—IMO Policies and Practices Related to the Reduction of Greenhouse Gas Emissions from Ships*. A 23/Res.963.

IMO. (2015). *Third IMO Greenhouse Gas Study 2014*. International Maritime Organization.

IMO. (2017, February 21). *Consideration of How to Progress the Matter of Reduction of GHG Emissions from Ships: Existing IMO Activity Related to Reducing GHG Emissions in the Shipping Sector—Note by the Secerariat*. ISWG-GHG1/2.

IMO. (2018a, April 16). *List of Participants*, MEPC 72/INF.1.

IMO (2018b, October 29). *List of Participants*, MEPC 73/INF.1.
IMO. (2019, May 21). *List of Participants*, MEPC 74/INF.1.
IMO. (2020). *Fourth IMO Greenhouse Gas Study 2020*. International Maritime Organization.
IMO. (2021, July 6). *List of Participants*, MEPC 76/INF.1.
IMO. (2022a). Energy Efficiency Measures. https://www.imo.org/en/OurWork/Environment/Pages/Technical-and-Operational-Measures.aspx. Last accessed: 20 March 2022.
IMO. (2022b). *Historic Background: IMO and the UNFCCC Framework*. https://www.imo.org/en/OurWork/Environment/Pages/Historic%20Background%20GHG.aspx. Last accessed: 19 March 2022.
IMO. (2022c). Structure of IMO. https://www.imo.org/en/About/Pages/Structure.aspx. Last accessed: 13 March 2022.
International Transport Forum. (2018). *Decarbonising Maritime Transport: Pathways to Zero-Carbon Shipping by 2035*. International Transport Forum Policy Papers, no. 47 OECD Publishing.
Juncos, A. E., & Pomorska, K. (2011). Invisible and Unaccountable? National Representative and Council Officials in EU Foreign Policy. *Journal of European Public Policy, 18*(8), 1096–1114.
Kassim, H., & Stevens, H. (2010). *Air Transport and the European Union: Europeanization and Its Limits*. Palgrave Macmillan.
Lee, J. W. (2017). A New Global Approach to Mitigating Emissions from International Aviation: The International Civil Aviation Organization's 39th Assembly (2016). *Chinese Journal of Environmental Law, 1*(1), 104–110.
Lin, W. (2017). Aviation and Climate Change: Practising Green Governmentality across the North-South Devide. *Geopolitics, 22*(1), 129–150.
Lindenthal, A. (2014). Aviation and Climate Protection: EU Leadership within the International Civil Aviation Organization. *Environmental Politics, 23*(6), 1064–1081.
Lyle, C. (2018). Beyond the ICAO's CORSIA: Towards a More Climatically Effective Strategy for Mitigation of Civil-Aviation Emissions. *Climate Law, 8*(1–2), 104–127.
Malina, R., McConnachie, D., Wichester, N., Wollersheim, C., Paltsev, S., & Waitz, I. A. (2012). The Impact of the European Union Emissions Trading Scheme on US Aviation. *Journal of Air Transport Management, 19*, 36–41.
Manners, I. (2002). Normative Power Europe: A Contradiction in Terms? *Journal of Common Market Studies, 40*(2), 235–258.
Martinez Romera, B. (2018). *Regime Interaction and Climate Change: The Case of International Aviation and Maritime Transport*. Routledge.
Martinez Romera, B., & van Asselt, H. (2015). The International Regulation of Aviation Emissions: Putting Differential Treatment into Practice. *Journal of Environmental Law, 27*(2), 259–283.

Minas, S., & Ntousas, V. (Eds.). (2018). *EU Climate Diplomacy: Politics, Law and Negotiations*. Routledge.

Nengye, L., & Maes, F. (2012). Legal Constraints to the European Union's Accession to the International Maritime Organization. *Journal of Maritime Law & Commerce, 43*(2), 279–291.

Oberthür, S. (2006). The Climate Change Regime: Interactions with ICAO, IMO, and the EU Burden-Sharing Agreement. In S. Oberthür & T. Gehring (Eds.), *Institutional Interaction in Global Environmental Governance Synergy and Conflict among International and EU Policies* (pp. 53–77). MIT Press.

Oberthür, S. (2003). Institutional Interaction to Address Greenhouse Gas Emissions from International Transport: ICAO, IMO and the Kyoto Protocol. *Climate Policy, 3*(3), 191–205.

Official Journal of the European Union. (2009). *Directive 2008/101/EC of the European Parliament and the Council of 19 November 2008 Amending Directive 2003/87/EC so as to Include Aviation Activities in the Scheme for Greenhouse Gas Emission Allowance Trading Within the Community (Text with EEA Relevance)*, L 8/3.

Official Journal of the European Union. (2015). *Regulation (EU) 2015/757 of the European Parliament and of the Council of 29 April 2015 on the Monitoring, Reporting and Verification of Carbon Dioxide Emissions from Maritime Transport, and Amending Directive 2009/16/EC (Text with EEA relevance)*, L 123/55.

Pace, M. (2007). The Construction of EU Normative Power. *Journal of Common Market Studies, 45*(5), 1041–1064.

Pallis, A. A. (2002). *The Common EU Maritime Transport Policy: Policy Europeanisation in the 1990s*. Routledge.

Penner, J. E., Lister, D. H., Griggs, D. J., Dokken, D. J., & McFarland, M. (Eds.). (1999). *Summary for Policy Makers: Aviation and the Global Atmosphere*. Intergovernmental Panel on Climate Change.

Petersen, M. (2008). The Legality of the EU's Stand-Alone Approach to the Climate Impact of Aviation: The Express Role Given to the ICAO by the Kyoto Protocol. *Review of European, Comparative & International Environmental Law, 17*(2), 196–204.

Pomorska, K., & Vanhoonacker, S. (2016). 'Europe as a Global Actor: Searching for a New Strategic Approach. *Journal of Common Market Studies, 54*(Annual Review), 204–217.

Preston, H., Lee, D. S., & Hooper, P. D. (2012). The Inclusion of the Aviation Sector within the European Union's Emissions Trading Scheme: What Are the Prospects for a More Sustainable Aviation Industry? *Environmental Development, 2*, 48–56.

Prum, D. A., & Kisska-Schulze, K. (2015). The Environmentally Conscious Skies: Did the European Union's Game of Brinksmanship Lead to a Viable

Global Plan for Emissions Trading in Aviation? *Washington University Global Studies Review, 14*(1), 1–48.
Rathbun, B. C. (2008). Interviewing and Qualitative Field Methods: Pragmatism and Practicalities. In J. M. Box Steffensmeier, H. E. Brady, & D. Collier (Eds.), *The Oxford Handbook of Political Methodology* (pp. 685–701). Oxford University Press.
Reagan, D. B. (2008). Putting International Aviation into the European Union Emissions Trading Scheme: Can Europe Do It Flying Solo? *Boston College Environmental Affairs Law, 35*(2), 349–384.
Ritchie, H. (2020, October 22). Climate Change and Flying: What Share of Global CO_2 Emissions Come from Aviation? *Our World Data*. https://ourworldindata.org/co2-emissions-from-aviation. Last accessed: 16 March 2022.
Rony, A. H., Kitada, M., Dalaklis, D., Ölçer, A. I., & Ballini, F. (2019). Exploring the New Policy Framework of Environmental Performance Management for Shipping: A Pilot Study. *WMU Journal of Maritime Affairs, 18*(1), 1–24.
Scott, J., & Rajamani, L. (2012). EU Climate Change Unilateralism. *European Journal of International Law, 23*(2), 469–494.
Shi, Y. (2018). The Implications of the Paris Agreement for the Regulation of Greenhouse Gas Emissions from International Shipping. *Ocean Yearbook, 32*, 528–555.
Shi, Y., & Gullet, W. (2018). International Regulation on Low-Carbon Shipping for Climate Change Mitigation: Development, Challenges and Prospects. *Ocean Development & International Law, 49*(2), 134–156.
Skjærseth, J. B., & Wettestad, J. (2008a). *EU Emissions Trading: Initiation, Decision-making and Implementation*. Ashgate.
Skjærseth, J. B., & Wettestad, J. (2008b). Implementing EU Emissions Trading: Success or Failure? *International Environmental Agreements: Politics, Law and Economics, 8*(3), 275–290.
Sochor, E. (1991). *The Politics of International Aviation*. Macmillan Press.
The Business Times. (2018, August 2). *EU an Avid Supporter of Rules-Based Global System*, Singapore. https://www.businesstimes.com.sg/opinion/eu-an-avid-supporter-of-rules-based-global-system. Last accessed: 13 March 2022.
The Guardian. (2021, October 14). UN Chief Urges Airlines and Shipping Firms to Do More to Cut Emissions. https://www.theguardian.com/environment/2021/oct/14/un-chief-urges-airlines-and-shipping-firms-to-do-more-to-cut-emissions?fbclid=IwAR2wlclWt2FajfrdrVHqj9OHDATumxNzZyAcyfcYYDfpav5MJYmCo2CnnAA. Last accessed: 13 March 2022.
Traut, M., Larkin, A., Anderson, K., McGlade, C., Sharmina, M., & Smith, T. (2018). CO_2 Abatement Goals for International Shipping. *Climate Policy, 18*(8), 1066–1075.

Van der Mei, A. P. (2016). Case Note: EU External Relations and Internal Inter-Institutional Conflicts: The Battlefield of Article 218 TFEU. *Maastricht Journal of European and Comparative Law, 23*(6), 1051–1076.
Van Evera, S. (1997). *Guide to Methods for Students of Political Science*. Cornell University Press.
Van Leeuwen, J. (2015). The Regionalization of Maritime Governance: Towards a Polycentric Governance System for Sustainable Shipping in the European Union. *Ocean & Coastal Management, 117*, 23–31.
Van Leeuwen, J., & Kern, K. (2013). The External Dimension of European Union Marine Governance: Institutional Interplay between the EU and the International Maritime Organization. *Global Environmental Politics, 13*(1), 69–87.
Vaugeois, S. (2019). International Civil Aviation Organization (ICAO): The Developments of the Position of the EU in International Aviation. In R. A. Wessel & J. Odermatt (Eds.), *Research Handbook on the European Union and International Organizations* (pp. 222–239). Edward Elgar Publishing.
Watanabe, R., & Robinson, G. (2005). The European Union Emissions Trading Scheme (EU ETS). *Climate Policy, 5*(1), 10–14.
Weber, L. (2021). *International Civil Aviation Organization (ICAO)* (4th ed.). Wolters Kluwer.
White, P. (1999). The Role of UN Specialised Agencies in Complex Emergencies: A Case Study of FAO. *Third World Quarterly, 20*(1), 223–238.
Wouters, J., de Jong, S., Marx, A., & De Man, P. (2009). *Study for the Assessment of the EU's Role in International Maritime Organizations*. Leuven Centre for Global Governance Studies, Katholieke Universiteit Leuven.
Zhang, H. (2016). Towards Global Green Shipping: The Development of International Regulation on Reduction of GHG Emissions from Ships. *International Environmental Agreements: Politics, Law and Economics, 16*(4), 561–577.

CHAPTER 2

Shaping the Field: Normative Power Europe, the EU Climate Norm, and EU (Climate) Diplomacy

1 Introduction

This chapter introduces the broader framework, the subjects, and the notions around which this book evolves. The next section (2) outlines the prominent theoretical framework of diffusing EU norms, namely Normative Power Europe; it also discusses alternative explanations to NPE, as the latter has not been widely applied to explain the EU's action towards other IOs. In the following section (3), the relation between NPE and the EU's climate diplomacy is unfolded; it begins with the evolution of tackling climate change as a distinct policy field in the European political arena and suggests that fighting climate change is indeed a European norm. It, then, reveals the distinct character of the EU's diplomacy within the broader framework of the policymaking of the EU's external action and continues by highlighting the saturation of the EU's climate diplomacy with normative characteristics. The final section (4) concludes the chapter and provides the way forward to the theoretical framework and the analysis of the book.

© The Author(s), under exclusive license to Springer Nature Switzerland AG 2024

G. Dikaios, *EU Climate Diplomacy Towards the IMO and ICAO*, Palgrave Studies in European Union Politics,
https://doi.org/10.1007/978-3-031-51123-3_2

2 Normative Power Europe (NPE): An Introduction

NPE is a theoretical approach that aims to outline the identity of the European Union in international politics, as well as to showcase how the EU acts in order to promote its norms externally. The first appearance of NPE in the literature was focused mainly on the protection of human rights and on the effort to establish normative power as a counterbalance to both military and civilian power (Manners, 2002, 2006a, 2006b, 2006c; Sjursen, 2006a, 2006b), after the end of the Cold War and the consequent changes in international cooperation (Manners, 2010). That endeavor can be viewed within the frame of the greater emphasis that had been given at the time to international institutions (Whitman, 2013; Zervaki, 2014) and to the protection of human rights (Manners, 2002; Youngs, 2004).

Since its conceptualization, NPE has received a spate of criticisms whose main target is the identification of a significant gap between the *EU's rhetoric* in promoting norms and values at the international level on one hand, and the actual *EU's practice* that seems to be rationally driven on the other (cf. Diez, 2004; Hyde-Price, 2006; Noutcheva, 2009; Roccu & Voltolini, 2018; Wood, 2009). Nevertheless, the term has been utilized in a critical manner to further the discussion on the EU's presence in the world, or the perception of the world through the theory itself (cf. Chaban et al., 2017; Del Sarto, 2016; Diez, 2013; Kavalski, 2013; Keene, 2012; Manners, 2013, 2015; Pace, 2007; Parker & Rosamond, 2013). The concept has been also applied to explain various phenomena in the external relations of the EU and the way that the Union employs norm diffusion. One of the most important such processes is that of democratic promotion, particularly in the European region, but also in the rest of the world (Laïdi, 2008; Neuman, 2019). At the same time, scholars have applied the basic assumptions of the NPE in other fields, such as trade (Behrens & Janusch, 2012; Leeg, 2014; Manners, 2009b), economic governance (Bengtsson, 2015), development (Birchfield, 2013), energy, and the environment and climate change (Afionis & Stringer, 2012, 2014; Falkner, 2007; Lightfoot & Burchell, 2005; Renckens et al., 2017).

The literature on NPE has evolved largely taking two strands. In the first, the normative *identity* of the Union in international relations is disputed on a theoretical level. In the second, on a more practical level, several mechanisms are considered, which explain how the Union *acts* when it exercises its normative power; this approach usually attempts to answer whether the EU is a normative power through the exploration of specific case studies. The divide emerges in Manners' 2002 article, which is separated into two parts: the first part discusses whether the EU can be conceptualized as a different kind of power, by focusing on comparing its innate characteristics with other types of power (i.e., military and civilian). The second part proposes several factors that can justify the unique characteristics of the EU's international action, which would allow one to claim that the EU is a different kind of power; the one that Manners named 'normative'. As put by Whitman, 'Manners distinguishes the two aspects of normative power: being normative and acting in a normative way' (Whitman, 2013: 176). Cutting across these two strands, another discussion further develops regarding the relation between norms and *strategic interests*, weighing the one against the other (Van Schaik & Schunz, 2012; Youngs, 2004).[1]

In examining the cases at hand (the EU–IMO and the EU–ICAO relations), as well as the diplomatic exercise of the EU towards the two IOs, the six factors that Manners developed in 2002 are employed, namely *contagion*, *informational diffusion*, *procedural diffusion*, *transference*, *overt diffusion*, and *cultural filter*, in order to explain the ways the EU externalizes its norms.[2] The implementation of these factors constitutes a comprehensive approach to the action of the Union when its goal is the modification of a given situation in a policy matter in accordance with the EU's values or practices. Scholars aiming to explain how the EU acts in its external relations applying the NPE approach, have mostly developed different factors or mechanisms, not limiting their research only on the ones introduced by Manners. Very similar to Manners' conceptualization are the factors developed by Forsberg (2011) who embeds mechanisms that have often been applied to explain the EU's external action in the NPE discourse. Forsberg proposes four main factors, namely *persuading others*, *invoking norms*, *shaping the discourse of what is normal*, and *the power of example*. Despite the different

[1] It is worth mentioning that an important criticism to NPE theory comes from the strategic and rational choice approaches. For example, Noutcheva claimed that the action of the EU (in Western Balkans) was rather rationally driven rather than normatively, cancelling the normative power justification in its core (Noutcheva, 2009).

[2] See also Table 3.3 herein for an extensive presentation on how these six factors function.

approaches, which we may encounter, that also act as proof that different mechanisms can be well integrated under NPE's mantle, the following section showcases the fact that other factors may easily fall within the scope of the six factors that Manners proposed in 2002, which have the ability to produce a comprehensive image and operate as umbrella mechanisms to shed light on the EU's action at the international level.

While some critics pay attention to the identity part of the NPE, critiques on each operational part are not abundant. Whichmann in 2010 in her book titled *Normative or Strategic Power Europe?* claimed that the normative power concept is not suitable for empirical analysis and that its operationalization is a very difficult path to follow (Whichmann, 2010). Whichmann's analysis, however, fails to address the ways in which Manners operationalized the term 'normative power Europe', while completely discards the application of his theoretical approach for the analysis of the EU's foreign policy. It is worth noting that, a few years earlier, Whichmann had written that, although scholars are very skeptical about whether the EU is a normative power, 'unfortunately most of the analysis stays on an abstract level without really looking at what the EU [actually] does' (Whichmann, 2007: 98). However, she herself then attempted, in her aforementioned book, to investigate what the EU does by using 'normative power Europe' theory, looking at *different* factors than those which the theory proposes and ending up rejecting it. As such, our approach is against her criticism by arguing that the initial operationalization of the NPE factors—that explain how the EU acts as a normative power—can sufficiently be applied to an empirical analysis of the EU's normative performance.

Alternative Explanations to NPE

In the external EU climate action literature, the NPE approach is not widely applied. The case is the same for the literature on the EU–IOs relations. Therefore, an overview of different analytical models that have been proposed to explore the EU's influence on multilateral organizations is important to be presented. A special focus is given on climate change politics/policies, the outcome of which justifies further the utilization of the NPE theoretical model.

The most prevailing approach attempting to analyze the EU's climate performance on the international stage is that of *leadership*. While present in the literature since the beginning of the 2000s (Gupta & Ringius, 2001), the *leadership* concept took a significant boost in scholarly discussion after the so-called failure at the Copenhagen Summit in 2009 (Bäckstrand & Elgström, 2013), shedding further light on the reasons of such failure. Scholars have developed a series of operationalizations in order to explore the role of the EU in international climate negotiations which, until today, is the most applied concept for that purpose (Groen et al., 2013; Kryżanowski, 2015; Torney, 2019).

Accordingly, one of the first important mentions of operationalizations of leadership is that of Wurzel and Connelly in 2011 who categorize the latter into four types, namely, the *structural*, the *entrepreneurial*, the *cognitive*, and the *symbolic*. In more detail, *structural* leadership derives from the hard power an actor has; *entrepreneurial* leadership emerges through the diplomatic efforts an actor puts as to conclude international agreements; *cognitive* leadership concerns the involvement of ideas in determining the possible outcomes; and *symbolic* leadership is the ostensible action of an actor towards achieving a goal, but without actually working towards it (Wurzel & Connelly, 2011: 13). According to Vogler, 'all are relevant to the analysis of the Union's role and its successes and failures' (Vogler, 2014: 82) and have been influential to the EU's international climate policy.

Various types of leadership have been described in an attempt to explain the EU's action in international climate negotiations. For example, in their work, Rayner and Jordan maintain structural leadership with the same meaning and propose three additional types of leadership, namely the *directional*, the *instrumental*, and the *idea based*.

> Directional leadership entails leading by example with bold unilateral policy making, in anticipation of a demonstration effect. Instrumental leadership involves the linkage of issues and coalition building to reach agreements [...]. Idea-based leadership, akin to the concept of entrepreneurship, entails exerting an impact on how international policy agendas are defined. (Rayner & Jordan, 2016)

In 2019, Jänicke and Wurzel, while maintaining the four factors that Wurzel and Connelly proposed back in 2011, discarded that of symbolic leadership and incorporated a slightly different type based on directional leadership, named exemplary leadership. This type of leadership 'occurs when actors intentionally or unintentionally set an example for others' (Jänicke & Wurzel, 2019: 23).

It has to be noted that the concept of leadership has, from early on, been associated with the normative power approach (Afionis & Stringer, 2012; Falkner, 2007). Leadership and normative power are not incompatible approaches but rather complementary ones. The choice of NPE over leadership is based on the fact that the NPE better allows exploring and scrutinizing the diplomatic practices and efforts by the EU officials towards the ICAO and the IMO. Although, the similarities with the leadership types are many, something that makes NPE a more attractive means to explore the EU's climate diplomacy, especially in contrast to leadership, is that it has not been systematically applied to explain this kind of action and that a more ample field for research exists, whereas leadership has been the major explanatory framework for over twenty years and could be seen as saturated.

The concept of leadership has a teleological inclination that does not assist in examining the performance of an entity macroscopically. Analyses that apply the concept of leadership usually examine specific case studies and provide positive or negative answers, being captivated to present an ephemeral image of the entity's performance.[3] In contrast, the normative character of activity by the EU needs time and continuous practice in order to be developed and to be effective. As a consequence, and since the EU—almost from the beginning of the development of its climate policy—had set some initial targets on mitigating GHG emissions by airplanes and ships (Oberthür, 2006), it seems more appropriate to explore its action towards the ICAO and the IMO through the lenses of a normative approach.

Similar criticism applies to another explanation theory that has been used to describe the performance of EU climate action, namely the *goal achievement* approach, which has also been intertwined with the concept of leadership (Groen et al., 2012). Goal achievement has been proposed by several scholars to measure the effectiveness of the EU in multilateral organizations and institutions (Jørgensen et al., 2011; Oberthür & Groen, 2015). This approach has been especially applied to multilateral environmental issues, such as biological diversity (Groen, 2019;

[3] A recent attempt to reverse this argument in support of leadership as systematic method of analyzing the EU's external climate action is coined by Oberthür and Dupont (2021).

Oberthür & Rabitz, 2014), forests (Delreux & Pirlot, 2017), but also arms trade (Romanyshyn, 2015).

Oberthür and Groen, in assessing the EU goal achievement in the Paris Agreement negotiations, came up with a three-level explanatory framework which consists of 'the EU's ability to achieve its goals', focusing on the power and interest relations in the international milieu, the 'domestic EU politics and policy' and 'the EU negotiation strategy and diplomacy' (Oberthür & Groen, 2018: 711–774). The first and main difference is that this model takes the assumption that the EU is able to play a role at the international level for granted, as it begins by examining the international environment rather than the internal EU context.

Another pitfall is that the goal achievement approach largely disregards the normative mechanisms that the EU can enable in order to influence a certain outcome, based possibly on a previous explanatory model on goal achievement that found that norms did not play a crucial role during the negotiations (Oberthür & Rabitz, 2014). Additionally, the model lacks focus on diplomacy. It concentrates largely on intra-EU elements (e.g., politics and strategies) and pays little attention to actual diplomatic practice. Very recent, and still limited, literature only addresses this issue by focusing on *diplomatic engagement* in pursuing goal achievement (Earsom & Delreux, 2021a; Groen, 2019; Oberthür & Dupont, 2021) and constitutes a fine starting point for the analysis of diplomatic efforts that the EU performs in the case studies at hand, where a more systematic analysis takes place. Goal achievement theory could be used as a complementary tool to the analytical framework of NPE and could be applied to explain how the EU possibly influenced the ICAO and the IMO to adopt ambitious climate policies, but its timeframe scope is much narrower (as it can be seen in the published analyses) and its focus on diplomacy has not had the chance to get into enough depth as of yet.

Other concepts have been also proposed, further than those of *leadership* and *goal achievement*, in order to explain the EU's influence over and within international organizations. One of the most common concepts that assesses EU foreign policy is that of *actorness*. Actorness interconnects four notions, namely recognition, authority, autonomy and cohesion; it has been systematically applied to the EU's relations with other international institutions. A common criticism is that it does not put enough emphasis on the external effects that the EU action has, while it also lacks operational concepts that would allow to measure the effectiveness of this action (Delreux, 2014; Gehring et al., 2013; Schunz and

Damro, 2020; Schunz et al., 2012). The concept of *inter-organizational influence* has sprung in the literature, concerning either the EU (Ewers-Peters, 2018) or climate change (Biedenkopf, 2017). Costa discusses the level of influence that an organization can impose on another, noticing two main analytical paths: 'first, one must measure the ability to influence' and 'second, the quality of the brought about by influence' (Costa, 2017: 394). Both approaches, especially the former, have been extensively applied in the literature and many more can be tracked down to their foundations. In order to be able to combine the two aforementioned concepts/streams, this of *performance*, which combines the previous two aforementioned ones will be used further down in the next chapter. One main reason for this choice is that, performance, because of its generic definition, allows for broader and at the same time more issue-specific conceptualizations. In a way, the notion of performance encompasses both actorness and inter-organizational influence, thus assisting the aim of this book in answering whether or not the EU performed normative climate diplomacy in a more holistic manner. Finally, the choice of the NPE approach as the main analytical tool is also based on the fact that little to no studies that concentrate on the EU's normative performance in multilateral organizations through the very concepts coined by Manners have been undertaken as yet.

3 NPE and EU Climate Diplomacy: Connecting the Dots

One of the most significant developments of the conceptualization of the EU as a normative power is the fact that a number of scholars link NPE and normative action to the internal coordination of interests and actions (Bengtsson, 2015; De Zutter, 2010; Sjursen, 2006a). This is due to the belief that if an internal norm is embedded in the functioning of the EU, it will be more robust and more reliable when it is promoted externally. The importance of the existence of a coherent block of the Member States towards combating environmental problems is also recognized by the European External Action Service (EEAS) which states that 'the external cannot be separated from the internal' (European Union, 2016: 17); this is also highlighted by the literature (Lavenex & Schimmelfenning, 2009; Van Schaik, 2010).

Another important element in the conceptualization of the normative external action of the EU, maybe less examined by the literature, is that of the concordance between the EU initiatives and the universally agreed or accepted legal norms (Sjursen, 2006a). Sjursen argues, in her article in which she questions whether the EU can be a normative power, that the EU action could be characterized by normative dimensions if it vows to strengthen the multilateral international order through international law and international institutions (Sjursen, 2006a). Bengtsson, enhancing the argument above, underlines the devotion of the EU to multilateralism by interconnecting it with (the) European 'value-based foreign policy' (Bengtsson, 2015: 39).

Therefore, it can be argued that fighting climate change is indeed a European norm which is externalized through the NPE approach, and for that purpose, a distinct apparatus has been developed within the EU for the *EU climate diplomacy*.

Fighting Climate Change: A European Norm

Is fighting climate change a European norm? More specifically, is the mitigation of GHG emissions a European norm? A norm in (world) politics is defined as 'a standard of appropriate behavior for actors with a given identity' (Finnemore & Sikkink, 1998: 891). Norms determine the actions of an actor which, first, will allow others to identify this actor as distinct and, second, will define this actor's behavior in pursuit of change according to its normative consignment (Katzenstein, 1996). For Manners (2021), addressing the challenges posed by climate change has emerged as one of the core norms that govern the actions and the behaviors of the EU. Nevertheless, there has not been a systematic exploration of this attribution to the EU's identity in the literature; it is usually taken for granted. This section assesses the climate action of the Union and argues in favor of the existence of the EU climate norm.

According to De Zutter in the article 'Normative Power Spotting: An Ontological and Methodological Appraisal', a norm needs to possess three core elements in order to be characterized as such. In her own words: 'Three interrelated sources are relevant for determining a normative power's norms: historical narratives; legal and institutional contexts; and internal and external practices' (De Zutter, 2010: 1111).

Climate change has been associated with GHG emissions.[4] This was clearly depicted in the UNFCCC in 1992, concentrating the efforts to deal with the challenge of climate change to be focused on the reduction of GHG emissions. To that end, major diplomatic efforts between states have been taking place ever since, in order to agree on the most sustainable solutions that would allow for economic development without further damaging the climate system.[5] At the same time, states began to develop climate policies to alleviate the possible effects of climate change.[6] In the European continent, in spite of the diverging policies assumed by several Member States in the beginning (Doussis, 2001), this role was assumed primarily by the European Union, which found itself in a position to have the knowledge and the resources to construct an integrative policy that can find application to its Member States.

European climate change policy was developed first and foremost as part of the broader European environmental policy (Vogler, 2014).[7] The defining characteristic of the EU environmental policy, which was not foreseen in the founding Treaty of Rome in 1957, was its evolvement on a functional basis, which to this day permeates the policy and its derivatives: the first common regulations for environmental protection in the late 1960s–early 1970s were drafted to secure the function of the Common Market and, consequently, to avoid distortions of competition. In the late 1980s, environmental protection was bound to economic and

[4] For the causal relationship between climate change and the GHG emissions see the five IPCC Assessment Reports (1990, 1995, 2001, 2007, 2014), which provide growing scientific evidence on the matter (Field et al., 2014; Houghton et al., 1990, 1995, 2001; Parry et al., 2007). An extensive interim report was also published by the IPCC in 2018 (Masson-Delmotte et al., 2018), pending the publication of the sixth Report.

[5] According to the UNFCCC, climate system 'means the totality of the atmosphere, hydrosphere, biosphere and geosphere and their interactions' (United Nations, 1992, article 1, para. 3).

[6] The UNFCCC is based upon the *principle of common but differentiated responsibility*, which translates to the different burden and actions that countries around the world have in order to address climate change challenges. Developed states are expected to carry the weight of action, while developing states are exempted from action until, supposedly, they achieve a certain level of development (United Nations, 1992, articles 3 and 4).

[7] For a thorough review of the history and the establishment of the EU environmental policy, see Avrami (2015), Jordan and Adelle (2013), Knill and Liefferink (2007).

social development,[8] while at the same period two principles were introduced and embedded in the Single European Act (1987) which cut across the EU environmental policy. First, the principle of integration, which places environmental protection at the core of the implementation of all the EU's policies (Doussis, 2009); second, the subsidiary principle, which delimits the involvement of the Union to the action that offers better results at the European level (Doussis, 2001). Thus, the EU's environmental policy—on the one hand—acquires a legal basis, something that assists in the alleviation of the tensions between the EU institutions and the Member States regarding who has the competence to take action, and—on the other—delineates the Union's action to the transboundary aspect of the environmental challenges. Further, the Maastricht Treaty in 1992 incorporated *sustainable development* as a European principle and raised environmental protection as one of the political priorities of the Union. The Treaty of Amsterdam in 1997 created a stronger institutional framework within which the decision-making was taking place in regard to the environmental priority.

Within this institutional framework, the configuration of the climate policy of the EU began, and since then runs, in parallel with the developments taking place at the international level. The major events in fighting climate change globally could be described as the following five: *First*, in 1988 the UN IPCC was established. Its goal is to inform the world about the newest scientific evidence on climate change and to assess the national and international policies with the target to mitigate GHG emissions (Doussis, 2020; Hulme & Mahony, 2010). *Second*, in 1992 the UNFCCC was adopted and entered into force in 1994, setting the foundational standards of the international cooperation in combating climate change. *Third*, the Kyoto Protocol to the UNFCCC was adopted in 1997 (it only entered into force in 2005), offering a first attempt and way/method towards mitigating GHG emissions by the developed states from 2008 through 2012. *Fourth*, the Copenhagen UNFCCC Summit took place in 2009; this Summit was supposed to deliver the subsequent scheme to the Kyoto Protocol but the negotiations failed miserably, and it instead produced a political declaration with no substantial measures, as a compromise between China and the United States of America (USA) at the eleventh hour (Whitelaw, 2009). *Fifth*, in 2015 the adoption of

[8] Following the global trend of the notion of sustainable development.

the Paris Agreement to the UNFCCC marked an important turn in international climate politics, uniting the whole world under obligations that strive towards tackling the challenges emanating from climate change.

In the EU, the origination of dealing with the concentration of GHG emissions in the atmosphere was launched in the mid-1980s with some occasional action by the European Parliament and the Commission. It was in 1990 that, during the June European Council, European leaders noted their concern over global warming, pledged to promote the adoption of an international climate agreement, and asked the Commission to propose measures for a European policy on reducing GHG emissions. Simultaneously, the need for a common position was advocated by the Council of the EU and through the Commission. In fact, the latter was being invited (initially as an observer) to the tables of international environmental negotiations[9] and was actively involved in internal and external EU climate change policymaking (Afionis, 2017; Schunz, 2014). Afionis (2017) claims further that the strive for an EU common position in the 'air' politics (ozone Montreal Protocol and UNFCCC) was a response, coined by European policymakers, to the failure of speaking with a single EU voice to other burning issues in the foreign and security policy fields of the time. Nevertheless, the evolution of the EU climate policy during its first decade was slow with no substantial actions taking place (Avrami, 2015; Delreux & Ohler, 2019).

The advancement of the EU climate policy came only after the signing of the Kyoto Protocol in 1997. While, during the negotiations of the Protocol, the EU had less influence on the final text in contrast to the USA, and deviated significantly from its starting position, the Union was considered as the broker of this bargain, for not leaving the table without a satisfying result. Apart from the several issues that appeared due to the simultaneous intra-EU consultations during the negotiations and the criticism that one could hold against the EU for various reasons (Afionis, 2017), it is worth mentioning on a positive note that, 'the EU for the first time, negotiated as a single bloc' (Delreux & Ohler, 2019: 12). This along with the USA distancing itself from the international climate regime, allowed the EU to open a new chapter in climate leadership and lead the way towards convincing third parties to implement the necessary measures to mitigate greenhouse gas emissions.

[9] For an extensive analysis on the EU negotiating nine multilateral environmental agreements, see Delreux (2011).

The leadership role was also acquired on the basis of the EU Emissions Trading System, which was introduced to the EU area in 2003 as the answer to the Kyoto Protocol's demands. The EU ETS 'is a cornerstone of the EU's policy to combat climate change and its key tool for reducing GHG emissions cost-effectively. [It posits] the EU [as] the world's first major carbon market and remains the biggest one' (European Commission, 2022). The EU ETS entered into force in 2005 and since then is in constant operation achieving (for several reasons and with a series of pitfalls) the set mitigation targets for the EU.[10]

Four years later, two significant incidents concerning EU climate policy occurred. The positive development was the entry into force of the Treaty of Lisbon on 1 December 2009. This was the first EU Treaty in which tackling climate change was explicitly mentioned, with a special note on the external scope of such efforts. This was not early enough, though, to forestall the negative, even disastrous, situation in the Copenhagen UNFCCC Summit that took place only a week after. The Summit symbolizes the abrupt ending of the grace period of the development of the EU's international relations, focusing on combating climate change. Not only were the EU positions profoundly ignored, but it was other countries with major interests in diminishing the importance of climate change, which negotiated the final output of the Summit without the EU.[11] The shock that the EU suffered led to a shift in how the UNFCCC goals would be achieved both internally and externally. Internally, a more comprehensive climate policy needed to be developed, independently of what was happening at the international level (Rayner & Jordan, 2016). Externally, the development of more active and engaging diplomatic relations was introduced aiming to secure a successful result in the possibly upcoming international agreement (Parker & Karlsson, 2017; Parker et al., 2017). In conjunction with the entry into force of the Treaty of Lisbon, which also resulted in the establishment of a new Directorate General on Climate

[10] For a comprehensive analysis on the EU ETS, see Skjærseth and Wettestad (2008a). For criticisms on its functioning, see Branger et al. (2015); Dirix et al. (2015). Revisions of the EU ETS Directive are frequent; the latest took place in 2018 and it is expected to be revised again to take into consideration the new targets under the European Green Deal.

[11] The EU was ostracized from the discussions with other major international players taking the lead and concluding to a disappointing result; the leading image of the EU was, thus, severely damaged. For analyses discussing the failure of the EU at the Copenhagen Summit, see Bäckstrand and Elgström (2013), Oberthür (2011).

Change (DG CLIMA), this was the first time that the term climate diplomacy gained ground in the European practice and in the literature on European Studies (Van Schaik, 2010).

European climate diplomacy successfully led to the conclusion of the Paris Agreement in 2015. This Agreement was seen as a great success of the EU as a whole—as well as of France's (Lütz et al., 2021; Pomorska & Vanhoonacker, 2016: 213)—providing space to bolster moves, both internally and externally, towards combating climate change through mitigating GHG emissions, highlighting one more opportunity to show the intrinsic connection between the internal and the external aspects of the European climate politics.[12]

The pole star of the EU strategy to implement the Paris Agreement, the European Green Deal, was devised in the following years and was presented in 2019 by the then-incoming President of the European Commission, von der Leyen. European Green Deal's main goal is to achieve climate neutrality by 2050 through the detachment of economic progress from the use of natural resources, a move which will allow for net zero GHG emissions (European Commission, 2019; cf. Eckert, 2021). Based on this seminal vision, the *European Climate Law* (Regulation (EU) 2021/1119) entered into force in July 2021 to translate the political goal of climate neutrality into a legal obligation (Doussis, 2021), validating the well-established commitment of the EU in mitigating GHG emissions, along with other functions serving this goal. The European Green Deal further confirms the international role that the EU has and must continue to have for ensuring that global efforts are in place, thus stressing the value of climate diplomacy in its toolbox.

Taking the current *status quo* into account, it is evident that the EU's climate policy is an evolving process and has a long-standing presence in

[12] The academic articles on European climate politics often use the *internal–external* division (Doussis, 2019; Dupont & Oberthür, 2015; Dupont et al., 2018; Minas & Ntousas, 2018; Oberthür & Pallemaerts, 2010a; Vogler, 2014), a division adopted by this book as well, when discussing the internal and external diplomatic aspects of the European action in international climate politics. It also has to be noted that there are numerous other internal and external climate change (and environmental) target examples that can be found in other EU policy sectors, further strengthening, thus, the claim that the fight against climate change is indeed a European norm. Indicatively, the EU Accession Agreements always have a chapter dedicated to environmental and climate targets, which the states that wish to become EU Members have to fulfill. It is supposed to be one of the most challenging goals to achieve, as the requirements are many and of a highly technical character (Sida's Helpdesk for Environment and Climate Change, 2012).

the European sphere. Without a doubt, the addendum of the external environmental competence with emphasis on the fight against climate change in the Treaty of Lisbon (2009) did, rather emphatically, reinforce the fact that the EU is committed to a *good cause*[13] for which it will be a pioneer at the international and global level.

Thus, through the consistent engagement with tackling climate change, the legal and institutional frameworks that have been developed throughout a period of over 30 years, and by vigorous internal and external action, fighting against climate change, with a special focus on reducing GHG emissions, can be considered as a distinct European norm; *the EU climate norm*.

It is worth mentioning that very few pieces of literature exist that make a direct reference to the EU's climate norm(s). An article that uses the term 'EU climate norms' is that of Braun (2014). The use of the plural in the term indicates that several EU norms are related to climate change. As a matter of fact, various scholars make reference to a series of climate norms that the EU projects at the international level. The most relevant to this study is the '2 °C target' (Gippner, 2016), accompanied by the 'commitment towards international law' and 'multilateralism' (Scheipers & Sicurelli, 2007); *norms* identified during the Kyoto Protocol negotiations. *Multilateralism* appears to be one of the most frequently appearing norms when the EU promotes climate rules (Braun, 2014; Van Schaik & Schunz, 2012), further supporting the argument for the study at hand that attempts to explain how the EU effectively influences other multilateral organizations. Interestingly enough, all the articles mentioned in this paragraph (except Gippner's) have the seminal article of Manners' on NPE as a starting point,[14] intertwining further the external action of the EU with its internal foundational principles.

What, further, is extracted from this section is that the EU's internal climate policy is heavily dependent on both external EU climate ambitions, which often lead the way for the international climate politics, and on the global political dynamics in general. As noted by some European technocrats: 'This dynamic of leadership and learning reflects the EU's

[13] In the June 1990 European Council conclusions, it is stated that 'the community must use more effectively its position of *moral* [emphasis given], economic and political authority...' (European Council, 1990).

[14] It is even more interesting that Gippner (2016) applies a norm diffusion approach without reference to NPE.

strong commitment to a multilateral response to climate change. It also reflects the EU's understanding that it cannot successfully combat climate change on its own' (Lefevere et al., 2015: 109). To that end, understanding the diplomatic system that the EU devises in order to convince others to assume (climate) action or to follow its example, is of utmost importance for the EU's external affairs.

Explaining Policy Making in Internal EU Diplomacy: A Starting Point

The entry into force of the Treaty of Lisbon and the increased supranational characteristics that were attributed to the external action and foreign policy of the Union widely opened the door to academic discussions on how the EU's strategies and objectives are being set up regarding its external affairs.

Thomas and Tonra summarized, in 2012, seven distinct models to explore, analyze, and explain 'the internal diplomacy of determining the EU's diplomatic objectives' (Thomas & Tonra, 2012: 13). The approaches that they include originate from different theoretical backgrounds and refer to scholars researching on EU foreign policy regardless of their theoretical nascence. More specifically, they classify their model into three broad theoretical categorizations, i.e., *intergovernmentalism*, *normative institutionalism*, and *constructivism-sociological institutionalism*. Table 1 summarizes the seven models:

For the cases at hand, the models that regard the intergovernmental approach fall out of the scope. Conversely, the approaches originating from the normative institutionalism approach[15] and those from constructivism-sociological institutionalism combine what has been discussed earlier and can be considered as the starting points in the endeavor to explain how the EU works in order to export its norms. In the case of combating climate change through mitigating GHG emissions, which, as discussed earlier, is an EU norm that the Union continuously attempts to externalize, the normative entrapment seems to fit fine with the argumentation of this book. Furthermore, as Thomas and Tonra

[15] For a comprehensive analysis on why applying normative institutionalism to explain EU foreign action, see Schimmelfennig and Thomas (2009) and Thomas (2009).

Table 1 Models for exploring, analyzing, and explaining the internal EU diplomacy

Intergovernmentalism	
Model 1: Competitive bargaining	Unanimous decisions
Model 2: Log-rolling	Decisions based on reciprocity
Normative institutionalism	
Model 3: Cooperative bargaining	Emphasis on the appearance of the EU as a global actor through joint action, 'consistency and coherence in EU policymaking across time and issue areas'
Model 4: Normative entrapment	Acting according to the EU norms and policy commitments, despite the policy preferences of Member States
Constructivism-sociological institutionalism	
Model 5: 'Brusselization'	Strengthening the power of the European institutions in order to make decisions to secure coherence in the external action
Model 6: Europeanization	Learning and adapting to the European norms that vest EU's foreign action. These norms are concurrently shaped through a bottom-up process informing the EU's institutions and the EU Member States' action in a two-level interaction
Model 7: Collective identity formation	Developing a common perception of the EU as a single space (not necessarily with a common identity)

Source Summarized by Thomas and Tonra (2012)

observe, the role the supranational European institutions have in this model is to 'frame and publicise policy choices in a manner that promotes agreements that are consistent with prior normative and policy commitments' (Thomas & Tonra, 2012: 20). Thus, the role of the European Commission to coordinate the EU Member States in order to culminate in a common diplomatic strategy is further highlighted, leading to what Thomas and Tonra call 'Brusselization'. Finally, it has to be mentioned that while the two authors observe these models as distinct starting points in analytical mindsets, what emerges is that some of them can work concurrently and explain different levels of activity during internal EU diplomacy.

The Distinct Character of EU Diplomacy
Keukeleire, in 2003, published an article that opened up the discussion about the distinct character that EU diplomacy presents in contrast to traditional diplomacy. Keukeleire (2003) splits the study of EU diplomacy into three strands. The first focuses on the evolution of the *Common Foreign and Security Policy*; the second focuses on the ***internal diplomacy*** that allows for common strategies to be adopted; and the third focuses on the attempts of the EU to mold policy changes in third parties. He names the latter strand ***structural diplomacy*** and proposes two main tools for its function: economic–financial instruments and political dialogue. At the same time, he argues that the EU performs structural diplomacy in policies falling under what was known as the first pillar.

The literature on EU diplomacy has mostly grown towards these three directions, with the first two usually being considered together in the literature on the development of the EEAS (e.g., Adler-Nissen, 2014; Baltag & Smith, 2015; Juncos & Pomorska, 2020; Spence & Bátora, 2015; Vanhoonacker & Pomorska, 2013). At the same time, much attention has been given to domestic politics and the internal diplomacy of the EU, as a crucial factor in performing effective external diplomacy (see the references above; Guggenbühl, 2013; Jørgensen, 2009). Furthermore, the analytical framework of structural diplomacy diverged from the Keukeleire's conceptualization (2003) as European diplomacy became more active, showcasing 'the process of dialogue and negotiation' with an ultimate goal of the objectives of this process 'to have an effect on structures and […] to have effects that are sustainable' (Keukeleire et al., 2009: 146). Some years later, the challenge of 'learning' also entered the discourse around structural diplomacy, broadening the scope of diplomatic functions (Keukeleire et al., 2016). According to the scholars cited, EU diplomacy offers a different point of view on what traditional diplomacy is and how it works. This comes as a result of the multidimensional character that diplomacy propounds in the post-Cold War era. Moreover, the connection between structural diplomacy (and its advancement) and the NPE approach becomes apparent, as the goal of the NPE, and in general of normative approaches, is to change the 'normal' of the receiver of the influence (Manners, 2002).

It is interesting to note that in the literature of European studies, the normative approach emanating from the work of Manners has not penetrated the study of the function of European diplomacy. The most notable, if not the only piece of literature is a book chapter by Manners and Whitman, that discusses the connection between 'normative power and the future of EU public diplomacy' (Manners & Whitman, 2013) and presents the ways in which the EU can apply the six factors that Manners proposed in 2002 with the aim to exert European norms. In their chapter, the authors note that 'what is necessary is an understanding of the specific mechanisms or modes of diffusion in the normative power approach and the roles they might play in EU public diplomacy' (Manners & Whitman, 2013: 184). At the same time, while its relation to the diplomacy field has not yet been carefully studied, normative power is one of the most widely applied approaches to European foreign policy in empirical studies (cf. Laïdi, 2008; Tocci, 2007, 2008). This divergence functions as the bridge for examining the EU's diplomatic efforts to convince the ICAO and the IMO to adopt ambitious climate targets through the normative perspective proposed by Manners.

EU Climate Diplomacy
Fighting climate change has been one of the priorities of the external action of the EU and several studies have pointed out that climate change can be considered as a stand-alone field within European foreign policy (Schunz, 2009, 2014; Schunz et al., 2009; Van Schaik & Schunz, 2012; Vogler, 2009).[16] Although the issues that concern the topics of climate diplomacy have been present for quite some time, the literature dealing with EU climate diplomacy per se and as a specific issue has flourished only over the past few years. The reasons for this shift to more direct scholarly mentions of the term are mainly two: first, as mentioned earlier, the Treaty of Lisbon explicitly referred to the efforts that the EU needs to put in combating climate change through international negotiations; second, after the failure of the Copenhagen Summit, the EU reorganized its approach in dealing with climate diplomacy (Biedenkopf & Petri, 2021), 'broadening [as well] its epistemic community of climate diplomats' (Cross, 2018: 572).[17]

[16] Ott highlighted already from the beginning of the twenty-first century that climate change is 'an important foreign policy issue' (Ott, 2001). Moreover, Schunz et al. (2009) use extensively the term climate diplomacy.

[17] The literature review of this part is based on papers that focus on EU action and use the term *climate diplomacy*, at least once in their main body. Of course, there are numerous articles that analyze non-EU climate diplomacy aspects using the term *climate*

One of the first pieces of literature that employed the term *climate diplomacy* was a book that was edited by Oberthür and Pallemaerts in 2010. The book consists of an extensive chapter penned by Van Schaik concerning the evolution of the EU diplomatic system in regard to international climate negotiations towards the UNFCCC and, mostly, the internal dynamics of the Union in regard to the subject at hand (Oberthür & Pallemaerts, 2010b; Van Schaik, 2010). Since then, several studies have been published, proposing an array of different models on how to assess, and what the prospects are of, the role of the EU in climate negotiations. However, most of this literature has so far neglected to assess the role of the EU in a normative power approach. This lack of specific research based on the theory of normative power could be attributed to an array of reasons, including the following. The prevailing focus of the NPE has been in examining whether the identity of the Union is itself normative; the initial special focus of the conceptualization of NPE which gave emphasis to democratization and human rights (especially to third states and not to other multilateral organizations); and, the up until now, lack of a systematic reading of the fight against climate change as a core EU norm.

To this end, a prevalent strand of scientific publications is focusing on bilateral relations between the major actors in international climate negotiations (i.e., the EU, the USA, and China). This strand looks into the institutional pressures and challenges that the EU experiences (Bo et al., 2016; Torney, 2014), the interplay between the actors involved, and the new power politics that emerge in the new climate diplomacy (Belis et al., 2015). Further, it examines the way in which the EU climate norms infiltrate to third parties (Gippner, 2016). In addition, some studies do exist that focus on different bilateral relations, such as the one on the institutional challenges in advancing European climate diplomacy towards the Association of Southeast Asian Nations (known as ASEAN) (Goron, 2014).

Country-specific analyses have also been present in the literature, with an emphasis, initially, on Germany and on its role in spreading the ambitions of the EU climate diplomacy (Frick & Schulz, 2014; Tänzler & Carius, 2013), towards Poland in an attempt to explain the skepticism towards those ambitions (Nowak, 2014) and, more recently, on

diplomacy (see, indicatively, DeCanio & Fremstad, 2013; Doussis, 2020; Korppoo, 2020; Light, 2017; Mabey et al., 2013).

the United Kingdom and on the possible negative effects that Brexit will have on EU's international climate policy and diplomacy (Bocse, 2020; Dupont & Moore, 2019). At the same time, studies offering a more general overview of the EU climate diplomacy have been published (e.g., Haug & Berkhout, 2010; Maciej Kaczyński, 2014; Schunz, 2015; Tänzler & Carius, 2013), complemented by more on the perceptions of EU climate diplomacy (Elgström, 2016), as well as on the way forward geopolitically wise (Oberthür, 2016).

Another strand in the literature of EU climate diplomacy worth mentioning is the study on the role of the various European institutions that promote the EU's climate targets. For example, the European Parliament plays a moderate role on the subject, while, at the same time, it is involved in international climate negotiations that aim to contribute to positive—and ambitious—outcomes (Delreux & Burns, 2019). Another EU body that is engaging in the subject is the EEAS. The EEAS—and its delegates around the world—are responsible to assist in making the EU's goals clearer, and to document that the action taken is evaluated as effective (Biedenkopf & Petri, 2019, 2021). The aforementioned contributions may be classified under the literature engaging with the EU internal constellations and their influence on international climate negotiations (Fischer & Geden, 2015).

A more systematic approach to the EU's climate diplomacy has been developed by Oberthür and Groen. Their first reference to climate diplomacy can be found in 2014; the authors are connecting EU climate diplomacy with the performance of the European Union in international climate negotiations (Oberthür & Groen, 2014). Adding to that, in 2018 they published an explanatory framework proposing a three-level analysis for EU climate diplomacy, which aims to offer a comprehensive image of how the EU achieves its goals (Oberthür & Groen, 2018).[18] Freshly, Oberthür and Dupont took the proposal for the analysis of the EU climate diplomacy a step further, by arguing in favor of the need for a *grand strategy* that would enhance the EU's leadership in the field by incorporating the majority (if not all) the EU's function in its external climate affairs (Oberthür & Dupont, 2021).

Last but not least, the book edited by Minas and Ntousas gathers, in good part, a broad array of information on so far under-researched

[18] See also sub-section *Alternative explanations to NPE* herein.

subjects concerning EU climate diplomacy and offers a comprehensive image of and a number of useful starting points on the numerous stages of EU climate diplomacy. The book's main focus is on the negotiations of technological advancements, energy, and financial issues emanating from the Paris Agreement, covering a wide spectrum of internal and external EU politics, law, and diplomacy subjects (Minas & Ntousas, 2018). In the same vein, the book edited by Adelle, Biedenkopf, and Torney looks into different aspects of EU climate diplomacy in several of its chapters, focusing more on the politics of climate diplomacy and on the relations between the EU and third parties (Adelle et al., 2018).

All this literature points to the direction that the EU has developed its climate diplomacy as a separate policy field. Adding to the discussion on the distinct character of EU diplomacy in general, the argument made by several scholars working on the concept, namely that the term *diplomacy* is today in flux (Adler-Nissen, 2014; Keukeleire et al., 2009; Petrov et al., 2012), seems to have a sound basis. Therefore, a need presents itself to define diplomacy in a solid manner in order to measure its effectiveness and analyze its outputs and outcomes. Petrov et al. (2012) suggest that *three aspects* challenge the traditional notion of diplomacy. The first concerns the inclusion of new policy areas in diplomatic processes; the second, the stronger interrelation between domestic and foreign policies; and the third, the proliferation of diplomatic actors. In the case of EU climate diplomacy, all three aspects are present. Climate diplomacy as a distinct field of foreign affairs has gradually started to emerge over the past years; climate diplomacy is heavily dependent on the domestic implementation of climate change policies; and the EU, the IMO, and the ICAO have been upgraded to core diplomatic actors, moving away from the old-fashioned trend which only considers states negotiating with each other. These observations also come in concordance with a new point of view that wants multiple actors playing the game with similar, or enough, power to influence the outcomes, in contrast with the Westphalian logic of a *dominant* actor, opening the way to different explanations of the way a system works (Belis et al., 2015).

It is evident from what was presented that there is a need to go beyond analyzing climate diplomacy and international climate politics through the notion of leadership (similar to *dominance* in the sense of an actor leading the way, influencing and ruling the system). Alternative analytical tools, such as normative power factors, need to be employed in order to acquire a sound understanding of the situation. To this end, Sjursen (2015), after all, supports that normative approaches in European foreign action literature have not yet been extensively (or systematically) applied. Further, Van Schaik and Shunz, in an article essential in conceptualizing the EU's external climate policy, note that 'the way normative considerations actually inform the Union's strategy for the international climate negotiations has been under-researched' (Van Schaik & Schunz, 2012: 170). Over ten years later, and despite the many academic pieces written on international climate politics, the normative justification of the EU strategies in international climate negotiations remains under-explored.

As mentioned earlier, the EU seems to have strategically decided to actively engage in international climate negotiations. Van Schaik and Schunz mention in the same piece of literature that 'social reality norms and interests can -and often do- coexist as determinants of actors' policy choices' (Van Schaik & Schunz, 2012: 171). To this extent, the need to view the actions of the EU and its overall performance that aims to pursue its interests through the lens of a normatively justified prism becomes evident.

Furthermore, one more aspect of the connection between normative power and climate diplomacy needs to be addressed. The European institution that has been the frontrunner in promoting the agenda of tackling climate change both at the EU and the international level, is undeniably the European Commission (Biedenkopf & Petri, 2021). The Commission's proactiveness and entrepreneurship have been reported in several cases and studies, with notable examples being these of the coordination processes which were conducted to reach the goal that the Kyoto Protocol stipulated.[19] Moreover, the addition of the EU ETS in the European political life was a success of the Commission's efforts, which went through fire and water in order to convince all EU Member States that it is beneficial for all, as it was essential for the goal of achieving the reduction of GHG emissions. Last but not least, the initiatives that the Commission implements regarding the mitigation of CO_2 emissions are pioneering and significantly more advanced in contrast to the policies of the Member States and, occasionally, even to the ones proposed at the international level (Rayner & Jordan, 2016). The logic behind these observations lies in the 'hesitant supranational turn' in policies

[19] The role of the Commission during climate negotiations, even at the beginning when it was informal, has been cited as instrumental in coordinating the EU Member States to appear united (Delreux, 2011); though not always successfully (Afionis, 2017).

mitigating GHG emissions, reported in the literature since the early 2010s (Wettestad et al., 2012); according to both the literature and the ways the Commission attempts to expand its influence regarding climate change policies in the previous years, this trend accelerates continuously (Nekvasil & Moldan, 2018). The final act of this play is the introduction of the European Green Deal and the adoption of the European Climate Law (Eckert, 2021) which significantly enhances the role of the Commission both internally and externally.

Another aspect that could be added to the discussion on the importance of the NPE approach for EU climate diplomacy is that the former can be viewed as inherently connected to the supranational character of the Union. First of all, it is the discussion on the identity of the EU as a whole that allows for strong normative action, based on pan-European values and principles. This is evident in an array of works concerning the EU's normative power, arguing in favor of the important role of coordination and coherence in the Member States' and EU institutions' actions in order not to deviate from the EU normative principles (e.g., Manners, 2006a; Wood, 2009). To this task, the role of the Commission, as the pre-eminent supranational institution, is deemed crucial, both in securing the normative character of the Union and in the strengthening of the latter where possible (Manners, 2009a, 2015; Noutcheva, 2008).

Adding to the above discussion the findings of studies which claim that the Commission influences the decision-making processes—mainly through its expertise—even to policies that it has no explicit competence over (e.g., EU foreign policy) (Chou & Riddervold, 2015; Riddervold, 2016; Riddervold & Rosén, 2015, 2016), it becomes unavoidable to concentrate in revealing the institutional constellations that allow (or impede) the Commission to play a significant role in shaping the negotiation strategies of the EU Member States. This aspect of the analysis emerges because of two noticeable characteristics that the cases under examination (climate change and transport) present.

Firstly, although both climate change and transport policies fall under the shared EU Member States competences, during the initial round of interviews, confusion was identified between the interviewees in regard to the question of competence. While some officials argued that the EU does possess the competence to deal with climate targets in international transportation, there have been others who viewed the subject differently and stated that, while the EU does not have the competence, the

Member States still ought to follow the common position agreed in Brussels (Interviews no. 1, 4, 5, 7, 11, 12, 13, 21). Secondly, the EU as a whole is not a full member of the IMO and the ICAO and it only holds an observer status; this reality might possibly be the main reason behind the uncertainty regarding competence, as only the EU Member States may officially speak during the meetings of the two IOs. Consequently, in order to understand the extent of the supranational influence of the EU on setting the agenda and defining the strategies of the Member States, this unique hybrid situation that emerges needs to be examined, as to prove that there is a strengthening of the role of the supranational institutions ('Brusselization')—even if the latter do not have an official voice.

4 Conclusion and the Way Forward

The NPE approach is one of the fundamental theoretical approaches of exploring the quest for a distinct EU identity that acts as a separate pillar of power in the international arena. Despite the popularity of NPE in analyses and its utilization as an initial conceptualization, its contribution to the Union's external operational activity and the mechanisms it proposes for the study of the diffusion of norms is largely unrecognized. Particularly in terms of its dialectical relationship with diplomacy, the literature on the subject suffers significantly. The same applies to the connection of NPE to the EU's influence on other IOs (who are global norm-makers).

The current state of emergency posed by the challenges of climate change puts states in a constant struggle to deal with the consequences, making flagship actions to reduce GHG emissions from all areas of anthropogenic action a necessity. The EU as a whole, through its many years of activity in the field of combating climate change, has made the latter a separate norm of its function, which permeates and guides—sometimes more and sometimes less—the political direction both internally and externally. The culmination of this action and a confirmation of the existence of this norm can be found in the recent adoption of the European Climate Law that infiltrates all areas of EU politics and everyday life.

The existence of the European climate norm and the simultaneous rapid increase in the weight given by the EU to its climate diplomacy pave the way for a systematic investigation of those factors that lead to the externalization of its climate norm. The indivisible link between internal and external diplomatic practices, especially in areas that have not been adequately studied, such as that of European climate diplomacy and international transport regulatory bodies, calls for further exploration—especially now that a large part of the international community and the EU are called on to take effective action to reduce their GHG pollution. In addition, this interconnection arises from the very nature of an actor who wishes to function as a normative power, as the establishment and consolidation of the internal operation of a norm facilitates and strengthens its externalization at the same time.

The external climate change policy of the EU is mainly focused on what is happening at the UNFCCC, something that also is depicted in the literature.[20] While, at first glance, it seems logical that scientific labor is leaning heavily towards the UNFCCC aspect, it is equally important to present that the EU attempts—for some years now—to also influence other regimes to adopt more climate-friendly targets. This becomes evident especially now, during a period that such activities seem to have matured, and reality points to the cross-sectoral aspect of fighting climate change in all levels of governance and different institutional settings (cf. Earsom & Delreux, 2021b).

To date, some scholars argue that the EU has failed to successfully project its normative power to climate change negotiations (Bäckstrand & Elgström, 2013; De Cock, 2011), while others showcase the positive influence that the EU normative power exerts on those negotiations (Gippner, 2016; Lightfoot & Burchell, 2005; Van Schaik & Schunz, 2012). Concurrently, articles related to EU–IMO (Nengye & Maes, 2010; Van Leeuwen & Kern, 2013) and EU–ICAO relations (Dejong, 2013; Lindenthal, 2014) place emphasis on the EU's international role and the institutional interplay and provide various angles of viewing how EU's policies and external action affect multilateral negotiations, without directly referencing the normative extensions of European influence. Further, as both the IMO and the ICAO were (and still are) reluctant to adopt ambitious but feasible targets to mitigate greenhouse gas emissions, the EU's diplomacy to pressure them to change their 'normal' business-as-usual plan and to become more open to the idea of, or even determined to, tackle climate change, requires consolidated research and exploration.

[20] References cited in sub-section *EU climate diplomacy* herein refer to the EU–UNFCCC relations.

List of Interviews

Interview no. 1, Advocate of the industry of an EU Member State, phone call, 31 March 2019.
Interview no. 4, EU Member State official, phone call, 16 April 2019.
Interview no. 5, EU Member State official, phone call, 16 April 2019.
Interview no. 7, EU Member State official, phone call, 25 April 2019.
Interview no. 11, EU Member State official, email exchanges, 30 April 2019.
Interview no. 12, EU Member State official, phone call, 2 May 2019.
Interview no. 13, EU Member State official, phone call, 3 May 2019.
Interview no. 21, EU Member State official, phone call, 22 May 2019.

References

Adelle, C., Biedenkopf, K., & Torney, D. (Eds.). (2018). *European Union External Environmental Policy: Rules, Regulation and Governance beyond Borders*. Palgrave Macmillan.

Adler-Nissen, R. (2014). Symbolic Power in European Diplomacy: The Struggle between National Foreign Services and the EU's External Action Service. *Review of International Studies, 40*(4), 657–681.

Afionis, S. (2017). *The European Union in International Climate Change Negotiations*. Routledge.

Afionis, S., & Stringer, L. (2012). European Union Leadership in Biofuels Regulation: Europe as a Normative Power? *Journal of Cleaner Production, 32*, 114–123.

Afionis, S., & Stringer, L. (2014). The Environment as a Strategic Priority in the European Union-Brazil Partnership: Is the EU Behaving as a Normative Power or Soft Imperialist? *International Environmental Agreements, 14*, 47–64.

Avrami, L. (2015). *Η πολιτική της Ευρωπαϊκής Ένωσης για την κλιματική αλλαγή: η συμμόρφωση ως παράγοντας διαμόρφωσης αποτελεσματικών εθνικών πολιτικών* [EU climate change policy: Compliance as a prerequisite for domestic policy effectiveness] [PhD Thesis]. National and Kapodistrian University of Athens. http://hdl.handle.net/10442/hedi/36213. Last accessed: 21 February 2021.

Bäckstrand, K., & Elgström, O. (2013). The EU's Role in Climate Change Negotiations: From Leader to 'Leadiator.' *Journal of European Public Policy, 20*(10), 1369–1386.

Baltag, D., & Smith, M. (2015). EU and Member State Diplomacies in Moldova and Ukraine: Examining EU Diplomatic Performance Post-Lisbon. *European Integration Online Papers*, Vol. 19, article 5.

Behrens, M., & Janusch, H. (2012). 'Great Normative Power': The European and American Trade Approach with Chile and Mexico. *European Foreign Affairs Review, 17*(3), 367–386.

Belis, D., Joffe, P., Kerremans, B., & Qi, Y. (2015). China, the United States and the European Union: Multiple Bilateralism and Prospects for a New Climate Change Diplomacy. *Climate Change Law Review, 3*, 203–218.

Bengtsson, R. (2015). The EU and Global Economic Governance: Playing the Role of a Global Leader? In A. Bjorkdahl et al. (Eds.), *Importing EU Norms: Conceptual Framework and Empirical Findings* (pp. 39–54). Springer.

Biedenkopf, K. (2017). Relations between International Organizations in Combating Climate Change. In R. Biermann & J. A. Koops (Eds.), *The Palgrave Handbook of Inter-organizational Relations in World Politics* (pp. 649–676). Palgrave Macmillan.

Biedenkopf, K., & Petri, F. (2019). EU Delegations in European Union Climate Diplomacy: The Role of Links to Brussels, Individuals and Country Contexts. *Journal of European Integration, 41*(1), 47–63.

Biedenkopf, K., & Petri, F. (2021). The European External Action Service and EU Climate Diplomacy: Coordinator and Supporter in Brussels and Beyond. *European Foreign Affairs Review, 26*(1), 71–86.

Birchfield, V. (2013). A Normative Power Europe Framework of Transnational Policy Formation. *Journal of European Public Policy, 20*(6), 907–922.

Bo, Y., Biedenkopf, K., & Chen, Z. (2016). Chinese and EU Climate and Energy Security Policy. In E. J. Kirchner, T. Christiansen, & H. Dorussen (Eds.), *Security Relations between China and the European Union: From Convergence to Cooperation?* (pp. 102–123). Cambridge University Press.

Bocse, A. M. (2020). The UK's Decision to Leave the European Union (Brexit) and Its Impact on the EU as a Climate Change Actor. *Climate Policy, 20*(2), 265–274.

Branger, F., Lecuyer, O., & Quirion, P. (2015). The European Union Emissions Trading Scheme: Should We Throw the Flagship Out with the Bathwater? *Wires Climate Change, 6*(1), 9–16.

Braun, M. (2014). EU Climate Norms in East-Central Europe. *Journal of Common Market Studies, 52*(3), 445–460.

Chaban, N., Knodt, M., & Verdun, A. (Guest Eds.). (2017). Special Issue: 'External images of the EU-Energy power Across the Globe'. *Comparative European Politics, 15*(1), 1–155.

Chou, M. H., & Riddervold, M. (2015). The Unexpected Negotiator at the Table: How the European Commission's Expertise Informs Intergovernmental EU Policies. *Politics and Governance, 3*(1), 61–72.

Costa, O. (2017). Assessing Influence between International Organizations. In R. Biermann & J. A. Koops (Eds.), *The Palgrave Handbook of Inter-organizational Relations in World Politics* (pp. 389–406). Palgrave Macmillan.

Cross, M. K. D. (2018). Partners at Paris? Climate Negotiations and Transatlantic Relations. *Journal of European Integration, 40*(5), 571–586.

De Cock, G. (2011). The European Union as a Bilateral 'Norm Leader' on Climate Change vis-à-vis China. *European Foreign Affairs Review, 16*(1), 89–105.

De Zutter, E. (2010). Normative Power Spotting: An Ontological and Methodological Appraisal. *Journal of European Public Policy, 17*(8), 1106–1127.

DeCanio, S. J., & Fremstad, A. (2013). Game Theory and Climate Diplomacy. *Ecological Economics, 85*, 177–187.

Dejong, S. M. (2013). Hot Air and Hot Heads: An Examination of the Legal Arguments Surrounding the Extension of the European Union's Emissions Trading Scheme to Aviation. *Asian Journal of International Law, 3*(1), 163–188.

Del Sarto, R. A. (2016). Normative Empire Europe: The European Union, Its Borderlands, and the 'Arab Spring.' *Journal of Common Market Studies, 54*(2), 215–232.

Delreux, T. (2011). *The EU as International Environmental Negotiator*. Ashgate.

Delreux, T. (2014). EU Actorness, Cohesiveness and Effectiveness in Environmental Affairs. *Journal of European Public Policy, 21*(7), 1017–1032.

Delreux, T., & Burns, C. (2019). Parliamentarizing a Politicized Policy: Understanding the Involvement of the European Parliament in UN Climate Negotiations. *Politics and Governance, 7*(3), 339–349.

Delreux, T., & Ohler, F. (2019). Climate Policy in the European Union Politics. *Oxford Research Encyclopedia: Politics*. https://doi.org/10.1093/acrefore/9780190228637.013.1097

Delreux, T., & Pirlot, P. (2017). The EU Performance in the UN Forum on Forests: An Analysis of the EU at UNFF11. In S. Blavoukos & D. Bourantonis (Eds.), *The EU in UN Politics: Actors, Processes and Performances* (pp. 187–207). Palgrave Macmillan.

Diez, T. (2004). Constructing the Self and Changing Others: Reconsidering NPE. *Millennium: Journal of International Studies, 33*(3), 613–636.

Diez, T. (2013). Normative Power as Hegemony. *Cooperation and Conflict, 48*(2), 194–210.

Dirix, J., Peeters, W., & Sterckx, S. (2015). Is the EU ETS a Just Climate Policy? *New Political Economy, 20*(5), 702–724.

Doussis, E. (2001). *Η Κοινοτική Πολιτική Περιβάλλοντος και η Επίδρασή της στην Περίπτωση της Ελλάδας* [The Community's Environmental Policy and Its Impact on the Greek Case]. Papazisi Publications.

Doussis, E. (2009). Le principle d'intégration. In P. Billet, M. Durousseau, G. Martin, & I. Trinquelle (Eds.), *Droit de l'environnement et protection de la santé* (pp. 255–260). L'Harmattan.

Doussis, E. (2019). Η Εξωτερική Δράση της ΕΕ για το Περιβάλλον και το Παράδειγμα της Κλιματικής Αλλαγής [The External Action of the EU on the Environment and the Example of Climate Change]. In S. Blavoukos, D. Bourantonis, & P. Tsakonas (Eds.), *Εξωτερικές Σχέσεις της ΕΕ* [The External Relational of the EU] (pp. 303–320). I. Sideris Publications.
Doussis, E. (2020). *Διεθνές Δίκαιο και Διπλωματία της Κλιματικής Αλλαγής* [International Law and Climate Change Diplomacy]. Nomiki Vivliothiki.
Doussis, E. (2021, 30 August–3 September). *The Road to Climate Neutrality and Just Transition in Southern Europe: The Case of Greece*. Paper presented at the ECPR General Conference, Virtual Event.
Dupont, C., & Moore, B. (2019). Brexit and the EU in Global Climate Governance. *Politics and Governance, 7*(3), 51–61.
Dupont, C., & Oberthür, S. (Eds.). (2015). *Decarbonization in the European Union: Internal Policies and External Strategies*. Palgrave Macmillan.
Dupont, C., Oberthür, S., & Biedenkopf, K. (2018). Climate Change: Adapting to Evolving Internal and External Dynamics. In C. Adelle, K. Biedenkopf, & D. Torney (Eds.), *European Union External Environmental Policy: Rules, Regulation and Governance Beyond Borders* (pp. 105–124). Palgrave Macmillan.
Earsom, J., & Delreux, T. (2021a). A Nice Tailwind: The EU's Goal Achievement at the IMO Initial Strategy. *Politics and Governance, 9*(3), 401–411.
Earsom, J., & Delreux, T. (2021b). Evaluating EU Responsiveness to the Evolution of the International Regime Complex on Climate Change. *International Environmental Agreements: Politics, Law and Economics, 21*(4), 711–728.
Eckert, S. (2021). The European Green Deal and the EU's Regulatory Power in Times of Crisis. *Journal of Common Market Studies, 59*(S1), 81–91.
Elgström, O. (2016). *Legitimacy, Credibility and Coherence—Perceptions of EU Roles in Global Climate Change Negotiations*. EUI Working Paper, RSCAS 2016/06.
European Commission. (2019). *Communication from the Commission to the European Parliament, the European Council, the Council, the European Economic and Social Committee and the Committee of the Regions: The European Green Deal*, COM(2019) 640 final.
European Commission. (2022). *EU Emissions Trading System (EU ETS)*. https://ec.europa.eu/clima/eu-action/eu-emissions-trading-system-eu-ets_en. Last accessed: 4 April 2022.
European Council. (1990). *Presidency Conclusions, Dublin, 25 and 26 June 1990*. SN 60/1/90.
European Union. (2016, June). *Shared Vision, Common Action: A Stronger Europe—A Global Strategy for the European Union's Foreign and*

Security Policy. http://www.eeas.europa.eu/archives/docs/top_stories/pdf/eugs_review_web.pdf. Last accessed: 13 March 2022.

Ewers-Peters, N. M. (2018). The EU as An Interorganizational Influencer? *European Foreign Affairs Review, 23*(3), 365–380.

Falkner, R. (2007). The Political Economy of 'Normative Power' Europe: EU Environmental Leadership in International Biotechnology Regulation. *Journal of European Public Policy, 14*(4), 507–526.

Field, C. B., Barros, V. R., Dokken, D. J., Mach, K. J., Mastrandrea, M. D., et al. (Eds.). (2014). *Climate Change 2014: Impacts, Adaptation, and Vulnerability*. Cambridge University Press.

Finnemore, M., & Sikkink, K. (1998). International Norm Dynamics and Political Change. *International Organization, 52*(4), 887–917.

Fischer, S., & Geden, O. (2015). The Changing Role in International Negotiations in EU Climate Policy. *The International Spectator: Italian Journal of International Affairs, 50*(1), 1–7.

Forsberg, T. (2011). Normative Power Europe, Once Again: A Conceptual Analysis of an Ideal Type. *Journal of Common Market Studies, 49*(6), 1183–1204.

Frick, M., & Schulz, S. (2014). Climate Change Politics and Policies: Germany and the European Union in the Climate Negotiations. In W. Hofmeister & P. Rueppel (Eds.), *Climate Change Diplomacy: The Way Forward for Asia and Europe* (pp. 37–52). Kondrad Adenauer Stiftung; East Asian Institute; European Union Centre in Singapore and Brussels: European Policy Centre.

Gehring, T., Oberthür, S. & Mühleck, M. (2013). 'European Union Actorness in International Institutions: Why the EU is Recognized as an Actor in Some International Institutions, but Not in Others'. *Journal of Common Market Studies, 51* (5), 849–865.

Gippner, O. (2016). The 2 °C Target: A European Norm Enters the International Stage-Following the Process to Adoption in China. *International Environmental Agreements, 16*, 49–65.

Goron, C. (2014). EU–ASEAN Relations in the Post-2015 Climate Regime: Exploring Pathways for Top-down and Bottom-up Climate Governance. In W. Hofmeister & P. Rueppel (Eds.), *Climate Change Diplomacy: The Way Forward for Asia and Europe*, pp. 101–129. Kondrad Adenauer Stiftung; East Asian Institute; European Union Centre in Singapore and Brussels: European Policy Centre.

Groen, L. (2019). Explaining European Union Effectiveness (Goal Achievement) in the Convention on Biological Diversity: The Importance of Diplomatic Engagement. *International Environmental Agreements: Politics, Law and Economics, 19*(1), 69–87.

Groen, L., Niemann, A., & Oberthür, S. (2012). The EU as A Global Leader? The Copenhagen and Cancun UN Climate Change Negotiations. *Journal of Contemporary European Research, 8*(2), 173–191.

Groen, L., Niemann, A., & Oberthür, S. (2013). The EU Role in International Climate Change Policy-Making: A Global Leader in Decline? In A. Boening, J. F. Kremer, & Van Loon, A. (Eds.) *Global Power Europe—Vol. 2: Policies, Actions and Influence of the EU's External Relations* (pp. 37–54). Springer.

Guggenbühl, A. (2013). The Culture of Negotiation in the European Union: Reviewing Trends and Predicting Patterns of Multilateral Decision-Making. *The Hague Journal of Diplomacy, 8*(1), 21–47.

Gupta, J., & Ringius, L. (2001). The EU's Climate Leadership: Reconciling Ambition and Reality. *International Environmental Agreements: Politics, Law and Economics, 1*(2), 281–299.

Haug, C., & Berkhout, F. (2010). Learning the Hard Way? European Climate Policy after Copenhagen. *Environment: Science and Policy for Sustainable Development, 52*(3), 20–27.

Houghton, J. T., Ding, Y., Griggs, D. J., Noguer, M., van der Linden, P. J., Dai, X., Maskell, K., & Johnson, C. A. (Eds.). (2001). *Climate Change 2001: The Scientific Basis*. Cambridge University Press.

Houghton, J. T., Jenkins, G. J., & Ephraums, J. J. (Eds.). (1990). *Climate Change: The IPCC Scientific Assessment*. Cambridge University Press.

Houghton, J. T., Meira Filho, L. G., Callander, B. A., Harris, N., Kattenberg, A., & Maskell, K. (Eds.). (1995). *Climate Change 1995: The Science of Climate Change*. Cambridge University Press.

Hulme, M., & Mahony, M. (2010). Climate Change: What Do We Know About the IPCC? *Progress in Physical Geography, 34*(5), 705–718.

Hyde-Price, A. (2006). 'Normative' Power Europe: A Realist Critique. *Journal of European Public Policy, 13*(2), 217–234.

Jänicke, M., & Wurzel, R. K. W. (2019). Leadership and Lesson-drawing in the European Union's Multilevel Climate Governance. *Environmental Politics, 28*(1), 22–42.

Jørgensen, K. E. (2009). The European Union in Multilateral Diplomacy. *The Hague Journal of Diplomacy, 4*(2), 189–200.

Jørgensen, K. E., Oberthür, S., & Shahin, J. (2011). Introduction: Assessing the EU's Performance in International Institutions—Conceptual Framework and Core Findings. *Journal of European Integration, 33*(6), 599–620.

Jordan, A., & Adelle, C. (Eds.). (2013). *Environmental Policy in the EU: Actors, Institutions and Processes* (3rd ed.). Routledge.

Juncos, A., & Pomorska, K. (2020). The European External Action Service. *Oxford Research Encyclopedias: Politics*. https://doi.org/10.1093/acrefore/9780190228637.013.1081

Katzenstein, P. J. (1996). Introduction: Alternative Perspectives on National Security. In P. J. Katzenstein (Ed.), *The Culture of National Security: Norms and Identity in World Politics* (pp. 1–32). Columbia University Press.

Kavalski, E. (2013). The Struggle for Recognition of Normative Power: Normative Power Europe and Normative Power China in Context. *Cooperation and Conflict, 48*(2), 247–267.

Keene, E. (2012). Social Status, Social Closure and the Idea of Europe as a 'Normative Power.' *European Journal of International Relations, 19*(4), 939–956.

Keukeleire, S. (2003). The European Union as a Diplomatic Actor: Internal, Traditional, and Structural Diplomacy. *Diplomacy and Statecraft, 14*(3), 31–56.

Keukeleire, S., Keuleers, F., & Raube, K. (2016). The EU, Structural Diplomacy and the Challenge of Learning. In M. Smith, S. Keukeleire, & S. Vanhoonacker (Eds.), *The Diplomatic System of the European Union: Evolution, Change and Challenges* (pp. 199–214). Routledge.

Keukeleire, S., Thiers, R., & Justaert, A. (2009). Reappraising Diplomacy: Structural Diplomacy and the Case of the European Union. *The Hague Journal of Diplomacy, 4*(2), 143–165.

Knill, C., & Liefferink, D. (2007). *Environmental Politics in the European Union: Policy-making, Implementation and Patterns of Multilevel Governance*. Manchester University Press.

Korppoo, A. (2020). Domestic Frames on Russia's Role in International Climate Diplomacy. *Climate Policy, 20*(1), 109–123.

Kryżanowski, M. (2015). International Leadership Re-/constructed? Ambivalence and Heterogeneity of Identity Discourses in European Union's Policy on Climate Change. *Journal of Language and Politics, 14*(1), 110–133.

Laïdi, Z. (Ed.). (2008). *EU Foreign Policy in a Globalized World: Normative Power and Social Preferences*. Routledge.

Lavenex, S., & Schimmelfenning, F. (2009). EU Rules Beyond EU Borders: Theorizing External Governance in European Politics. *Journal of European Public Policy, 16*(6), 791–812.

Leeg, T. (2014). Normative Power Europe? The European Union in the Negotiations on a Free Trade Agreement with India. *European Foreign Affairs Review, 19*(3), 335–356.

Lefevere, J., Runge Metzger, A., & Werksman, J. (2015). The EU and International Climate Change Policy. In J. Delbeke & P. Vis (Eds.), *EU Climate Policy Explained* (pp. 109–124). Routledge.

Light, A. (2017). Climate Diplomacy. In S. M. Gardiner & A. Thompson (Eds.), *The Oxford Handbook on Environmental Ethics* (pp. 487–500). Oxford University Press.

Lightfoot, S., & Burchell, J. (2005). The European Union and the World Summit on Sustainable Development: Normative Power Europe in Action? *Journal of Common Market Studies, 43*(1), 75–95.

Lindenthal, A. (2014). Aviation and Climate Protection: EU Leadership within the International Civil Aviation Organization. *Environmental Politics, 23*(6), 1064–1081.

Lütz, S., Leeg, T., Otto, D., & Woyames Dreher, V. (2021). *The European Union as A Global Actor: Trade, Finance and Climate Policy*. Springer.

Mabey, N., Gallagher, L., & Born, C. (2013). *Understanding Climate Diplomacy: Building Diplomatic Capacity and Systems to Avoid Dangerous Climate Change*. E3G Third Generation Environmentalism.

Maciej Kaczyński, P. (2014). EU Climate Diplomacy: Cautious Optimism Ahead of 2015. In W. Hofmeister & P. Rueppel (Eds.), *Climate Change Diplomacy: The Way Forward for Asia and Europe* (pp. 60–70). Kondrad Adenauer Stiftung; East Asian Institute; European Union Centre in Singapore and Brussels: European Policy Centre.

Manners, I. (2002). Normative Power Europe: A Contradiction in Terms? *Journal of Common Market Studies, 40*(2), 235–258.

Manners, I. (2006a). European Union 'Normative Power' and the Security Challenge. *European Security, 15*(4), 405–421.

Manners, I. (2006b). Normative Power Europe Reconsidered: Beyond the Crossroads. *Journal of European Public Policy, 13*(2), 182–199.

Manners, I. (2006c). The European Union as a Normative Power: A Response to Thomas Diez. *Millennium: Journal of International Studies, 35*(1), 167–180.

Manners, I. (2009a). The EU's Normative Power in Changing World Politics. In A. Gerrits (Ed.), *Normative Power Europe in A Changing World: A Discussion* (pp. 9–24). Netherlands Institute of International Relations Clingendael.

Manners, I. (2009b). The Social Dimension of EU Trade Policies: Reflection from a Normative Power Perspective. *European Foreign Affairs Review, 14*(5), 785–803.

Manners, I. (2010). Global Europa: Mythology of the European Union in World Politics. *Journal of Common Market Studies, 48*(1), 67–87.

Manners, I. (2013). Assessing the Decennial, Reassessing the Global: Understanding European Union Normative Power in Global Politics. *Cooperation and Conflict, 48*(2), 304–329.

Manners, I. (2015). Sociology of Knowledge and Production of Normative Power in the European Union's External Actions. *Journal of European Integration, 37*(2), 299–318.

Manners, I. (2021). Normative Power Approach to European Union External Action. In S. Gstöhl & S. Schunz (Eds.), *The External Action of the European Union: Concepts, Approaches, Theories* (pp. 69–84). Red Globe Press.

Manners, I., & Whitman, R. (2013). Normative Power and the Future of EU Public Diplomacy. In M. K. D. Cross & J. Melissen (Eds.), *European Public Diplomacy: Soft Power at Work* (pp. 183–204). Palgrave Macmillan.

Masson-Delmotte, V., Zhai, P., Pörtner, H. O., Roberts, D., Skea, J., Shukla, P. R., et al. (Eds.). (2018). *Global Warming of 1.5°C: An IPCC Special Report on the Impacts of Global Warming of 1.5°C Above Pre-Industrial Levels and Related Global Greenhouse Gas Emission Pathways, in the Context of Strengthening the Global Response to the Threat of Climate Change, Sustainable Development, and Efforts to Eradicate Poverty*. Intergovernmental Panel on Climate Change.

Minas, S., & Ntousas, V. (Eds.). (2018). *EU Climate Diplomacy: Politics, Law and Negotiations*. Routledge.

Nekvasil, M., & Moldan, B. (2018). Could Cohesion Policy Push EU Climate Efforts? *Climate Policy, 18*(1), 129–139.

Nengye, L., & Maes, F. (2010). The European Union and the International Maritime Organization: EU's External Influence on the Prevention of Vessel-Source Pollution. *Journal of Maritime Law & Commerce, 41*(4), 581–594.

Neuman, M. (Ed.). (2019). *Democracy Promotion and the Normative Power Europe Framework: The European Union in South Eastern Europe, Eastern Europe, and Central Asia*. Springer.

Noutcheva, G. (2008). Enlargement Policy towards Central and Eastern Europe, 1989–2007: Normative Intended. In N. Tocci (Ed.), *Who Is A Normative Foreign Policy Actor? The European Union and Its Global Partners* (pp. 26–31). Centre for European Policy Studies.

Noutcheva, G. (2009). Fake, Partial and Imposed Compliance: The Limits of the EU's Normative Power in the Western Balkans. *Journal of European Public Policy, 16*(7), 1065–1084.

Nowak, B. (2014). Domestic Sources of Multilateral Diplomacy—The Case of Poland's Climate Policy. In W. Hofmeister & P. Rueppel (Eds.), *Climate Change Diplomacy: The Way Forward for Asia and Europe* (pp. 53–59). Kondrad Adenauer Stiftung; East Asian Institute; European Union Centre in Singapore and Brussels: European Policy Centre.

Oberthür, S. (2006). The Climate Change Regime: Interactions with ICAO, IMO, and the EU Burden-Sharing Agreement. In S. Oberthür & T. Gehring (Eds.), *Institutional Interaction in Global Environmental Governance Synergy and Conflict Among International and EU Policies* (pp. 53–77). MIT Press.

Oberthür, S. (2011). The European Union's Performance in the International Climate Change Regime. *Journal of European Integration, 33*(6), 667–682.

Oberthür, S. (2016). Where to Go from Paris? The European Union in Climate Geopolitics. *Global Affairs, 2*(2), 119–130.

Oberthür, S., & Dupont, C. (2021). The European Union's International Climate Leadership: Towards A Grand Climate Strategy? *Journal of European Public Policy, 28*(7), 1095–1114.

Oberthür, S., & Groen, L. (2014). EU Performance in the International Climate Negotiations in 2013: Scope for Improvement. *Institute for European Studies*, Policy brief, iss. 01.

Oberthür, S., & Groen, L. (2015). The Effectiveness Dimension of the EU's Performance in International Institutions: Toward a More Comprehensive Assessment Framework. *Journal of Common Market Studies, 53*(6), 1319–1335.

Oberthür, S., & Groen, L. (2018). Explaining Goal Achievement in International Negotiations: The EU and the Paris Agreement on Climate Change. *Journal of European Public Policy, 25*(5), 708–727.

Oberthür, S., & Pallemaerts, M. (2010a). The EU's Internal and External Climate Policies: An Historical Overview. In S. Oberthür & M. Pallemaerts (Eds.), *The New Climate Policies of the European Union: Internal Legislation and Climate Diplomacy* (pp. 27–64). Brussels University Press.

Oberthür, S., & Pallemaerts, M. (Eds.). (2010b). *The New Climate Policies of the European Union: Internal Legislation and Climate Diplomacy*. Brussels University Press.

Oberthür, S., & Rabitz, F. (2014). On the EU's Performance and Leadership in Global Environmental Governance: The Case of the Nagoya Protocol. *Journal of European Public Policy, 21*(1), 39–57.

Ott, H. E. (2001). Climate Change: An Important Foreign Policy Issue. *International Affairs, 77*(2), 277–296.

Pace, M. (2007). The Construction of EU Normative Power. *Journal of Common Market Studies, 45*(5), 1041–1064.

Parker, C. F., & Karlsson, S. (2017). The European Union as A Global Climate Leader: Confronting Aspiration with Evidence. *International Environmental Agreements: Politics, Law and Economics, 17*(4), 445–461.

Parker, C. F., Karlsson, S., & Hjerpe, M. (2017). Assessing the European Union's Global Climate Change Leadership: From Copenhagen to the Paris Agreement. *Journal of European Integration, 39*(2), 239–252.

Parker, O., & Rosamond, B. (2013). NPE Meets Economic Liberalism: Complicating Cosmopolitanism Inside/Outside the EU. *Cooperation and Conflict, 48*(2), 229–246.

Parry, M., Canziani, O., Palutikof, J., van der Linden, P., & Hanson, C. (Eds.). (2007). *Climate Change 2007: Impacts, Adaptation and Vulnerability*. Cambridge University Press.

Petrov, P., Pomorska, K., & Vanhoonacker, S. (2012). The Emerging EU Diplomatic System: Opportunities and Challenges After Lisbon. *The Hague Journal of Diplomacy, 7*(1), 1–9.

Pomorska, K., & Vanhoonacker, S. (2016). 'Europe as a Global Actor: Searching for a New Strategic Approach. *Journal of Common Market Studies, 54*(Annual Review), 204–217.

Rayner, T., & Jordan, A. (2016). Climate Change Policy in the European Union. *Oxford Research Encyclopedia: Climate Science*. https://doi.org/10.1093/acrefore/9780190228620.013.47

Renckens, S., Skogstad, G., & Mondu, M. (2017). When Normative and Market Power Interact: The European Union and Global Biofuels Governance. *Journal of Common Market Studies, 55*(6), 1432–1448.

Riddervold, M. (2016). (Not) in the Hands of the Member States: How the European Commission Influences EU Security and Defence Policies. *Journal of Common Market Studies, 54*(2), 353–369.

Riddervold, M., & Rosén, G. (2015). Beyond Intergovernmental Cooperation: The Influence of the European Parliament and the Commission on EU Foreign and Security Policies. *European Foreign Affairs Review, 20*(3), 399–418.

Riddervold, M., & Rosén, G. (2016). Trick and Treat: How the Commission and the European Parliament Exert Influence in European Foreign and Security Policies. *Journal of European Integration, 38*(6), 687–702.

Roccu, R., & Voltolini, B. (2018). Framing and Reframing the EU's Engagement with Mediterranean: Examining the Security-Stability Nexus Before and After the Arab Uprisings. *Mediterranean Politics, 23*(1), 1–22.

Romanyshyn, I. (2015). Explaining EU Effectiveness in Multilateral Institutions: The Case of the Arms Trade Treaty Negotiations. *Journal of Common Market Studies, 53*(4), 875–892.

Scheipers, S., & Sicurelli, D. (2007). Normative Power Europe: A Credible Utopia? *Journal of Common Market Studies, 45*(2), 435–457.

Schimmelfennig, F., & Thomas, D. C. (2009). Normative Institutionalism and EU Foreign Policy in Comparative Perspective. *International Politics, 46*(4), 491–504.

Schunz, S. (2009). The European Union's External Climate Policy: A Foreign Policy Analysis. *CFSP Forum, 7*(2), 6.

Schunz, S. (2014). *European Foreign Policy and the Global Climate Regime*. P.I.E. Peter Lang.

Schunz, S. (2015). The European Union's Climate Change Diplomacy. In J. A. Koops & G. Macaj (Eds.), *The European Union as a Diplomatic Actor* (pp. 178–201). Palgrave Macmillan.

Schunz, S., Basu, S., Bruyninckx, H., Keukeleire, S. & Wouters, J. (2012). 'Analysing the Position of the European Union in the United Nations System: Analytical Framework'. In Wouters, J., Bruyninckx, H., Basu, S. & Schunz, S. (Eds.)., *The European Union and Multilateral Governance: Assessing EU*

Participation in the United Nations Human Rights and Environmental Fora (Basingstoke: Palgrave Macmillan), 25–48.

Schunz, S. & Damro, C. (2020). 'Expanding Actorness to Explain EU External Engagement in Originally Internal Policy Areas'. *Journal of European Public Policy, 27* (1), 122–140.

Schunz, S., Happaerts, S., & Van Den Brande, K. (2009). European Union Foreign Policy and Global Climate Change: Towards a Comprehensive European Climate Diplomacy?. *Leuven Centre for Global Governance Studies*, Policy Brief No. 12.

Sida's Helpdesk for Environment and Climate Change. (2012). *Western Balkan—Environment and Climate Change Policy Brief*. University of Gothenburg & SLU.

Sjursen, H. (2015). Normative Theory: An Untapped Resource in the Study of European Foreign Policy. In K. E. Jørgensen, Å. Kalland Aarstad, E. Drieskens, K. Laatikainen, & B. Tonra (Eds.), *The SAGE Handbook on European Foreign Policy* (Vol. 1, pp. 196–214). Sage.

Sjursen, H. (2006a). The EU as a 'Normative' Power: How Can This Be? *Journal of European Public Policy, 13*(2), 235–251.

Sjursen, H. (2006b). What Kind of Power? *Journal of European Public Policy, 13*(2), 169–181.

Skjærseth, J. B., & Wettestad, J. (2008a). *EU Emissions Trading: Initiation, Decision-making and Implementation*. Ashgate.

Skjærseth, J. B., & Wettestad, J. (2008b). Implementing EU Emissions Trading: Success or Failure? *International Environmental Agreements: Politics, Law and Economics, 8*(3), 275–290.

Spence, D., & Bátora, J. (Eds.). (2015). *The European External Action Service: European Diplomacy Post-Westphalia*. Palgrave Macmillan.

Tänzler, D., & Carius, A. (2013). Beyond International Climate Negotiations: Climate Diplomacy from a Foreign Policy Perspective. In O. C. Ruppel, C. Roschmann, & K. Ruppel Schlichting (Eds.), *Climate Change: International Law and Global Governance, Volume II: Policy, Diplomacy and Governance in A Changing Environment* (pp. 259–274). Nomos.

Thomas, D. C. (2009). Explaining the Negotiation of EU Foreign Policy: Normative Institutionalism and Alternative Explanations. *International Politics, 46*(4), 339–357.

Thomas, D. C., & Tonra, B. (2012). To What Ends EU Foreign Policy? Contending Approaches to the Union's Diplomatic Objectives and Representation. *The Hague Journal of Diplomacy, 7*(1), 11–29.

Tocci, N. (2007). *Profiling Normative Foreign Policy: The European Union and Its Global Partners*. CEPS Working Document, no. 279.

Tocci, N. (Ed.). (2008). *Who Is a Normative Foreign Policy Actor? The European Union and Its Global Partners*. Centre for European Policy Studies.

Torney, D. (2014). External Perceptions and EU Foreign Policy Effectiveness: The Case of Climate Change. *Journal of Common Market Studies, 52*(6), 1358–1373.

Torney, D. (2019). Follow the Leader? Conceptualising the Relationship between Leaders and Followers in Polycentric Climate Governance. *Environmental Politics, 28*(1), 167–186.

United Nations. (1992). *United Nations Framework Convention on Climate Change*.

Van Leeuwen, J., & Kern, K. (2013). The External Dimension of European Union Marine Governance: Institutional Interplay between the EU and the International Maritime Organization. *Global Environmental Politics, 13*(1), 69–87.

Van Schaik, L. (2010). The Sustainability of the EU's Model for Climate Diplomacy. In S. Oberthür & M. Pallemaerts (Eds.), *New Climate Policies of the European Union: Internal Legislation and Climate Diplomacy* (pp. 251–280). Brussels University Press.

Van Schaik, L., & Schunz, S. (2012). Explaining EU Activism and Impact in Global Climate Politics: Is the Union a Norm- or Interest-Driven Actor? *Journal of Common Market Studies, 50*(1), 169–186.

Vanhoonacker, S., & Pomorska, K. (2013). The European External Action Service and Agenda-setting in European Foreign Policy. *Journal of European Public Policy, 20*(9), 1316–1331.

Vogler, J. (2009). Climate Change and EU Foreign Policy: The Negotiation of Burden Sharing. *International Politics, 46*(4), 469–490.

Vogler, J. (2014). EU Climate Diplomacy—Internal and External Dimensions. In W. Hofmeister & P. Rueppel (Eds.), *Climate Change Diplomacy: The Way Forward for Asia and Europe* (pp. 71–88). Konrad Adenauer Stiftung; East Asian Institute; European Union Centre in Singapore and Brussels: European Policy Centre.

Wettestad, J., Eikeland, P. O., & Nilsson, M. (2012). EU Climate and Energy Policy: A Hesitant Supranational Turn? *Global Environmental Politics, 12*(2), 67–86.

Whichmann, N. (2007). Promoting the Rule of Law in the European Neighbourhood Policy—Strategic or Normative Power? *Politique Européenne, 2*(22), 81–104.

Whichmann, N. (2010). *Rule of Law Promotion in the European Neighbourhood Policy: Normative or Strategic Power?* Nomos.

Whitelaw, K. (2009, December 19). U.S., China Reach Tentative Climate Compromise. *NPR*. https://www.npr.org/templates/story/story.php?storyId=121600755&t=1649206689992. Last accessed: 31 May 2022.

Whitman, R. (2013). The Neo-Normative Turn in Theorizing the EU's International Presence. *Cooperation and Conflict, 48*(2), 171–193.

Wood, S. (2009). The European Union: A Normative or Normal Power? *European Foreign Affairs Review, 14*(2), 235–258.

Wurzel, R. K. W., & Connelly, J. (2011). Introduction: European Union Political Leadership in International Climate Change Politics. In R. K. W. Wurzel & J. Connelly (Eds.), *The European Union as a Leader in International Climate Change Politics* (pp. 3–20). Routledge.

Youngs, R. (2004). Normative Dynamics and Strategic Interests in the EU's External Identity. *Journal of Common Market Studies, 42*(2), 415–435.

Zervaki, A. (2014). *Resetting the Political Culture Agenda: From Polis to International Organization*. Springer.

CHAPTER 3

Performing Normative EU Climate Diplomacy: An Analytical Framework

1 Introduction

The objective of this chapter is to present the analytical framework and the work assumptions that guide the book to comprehensively study the EU's activity and action towards influencing the ICAO and the IMO in order to adopt rules that mitigate greenhouse gas emissions from aviation and shipping, respectively. The goal is to provide a holistic approach to the Union's performance from the moment a particular European policy appears on the agenda to the moment this policy concludes its aims. The main focus will be on the diplomatic efforts; in addition,—and in order to provide a clearer image—some aspects of their ramifications will be discussed. The purpose of this approach is to allow the empirical study of the EU's climate diplomacy towards the aforementioned international organizations. This holistic approach is separated into three distinct levels through which the theoretical assumptions that will attempt to explain the structural function of EU's operation are established: presentation and analysis of the internal procedures that take place at the European level; presentation and analysis of the external processes that the EU as a whole performs through the NPE approach; and tools for evaluation of the composite outcome of the two previous stages.

The structure of the present chapter is the following: Section 2 which immediately follows introduces the reasons behind the selection of the

© The Author(s), under exclusive license to Springer Nature Switzerland AG 2024

G. Dikaios, *EU Climate Diplomacy Towards the IMO and ICAO*, Palgrave Studies in European Union Politics,
https://doi.org/10.1007/978-3-031-51123-3_3

three-level analysis to explore the overall EU performance in the selected case studies. Section 3 develops the framework of analysis of the internal procedures and describes the relevant mechanisms used to examine the course of the EU's action, as well as brief literature reviews on each concept presented in order to establish their relevance to the study and the way they relate to each other. Section 4 presents the factors that NPE proposes for the study of externalizing norms. Section 5 unfolds the evaluation process of the EU performance. Section 6 underpins the analytical framework focusing on the discussion on what EU climate diplomacy is and clusters the theoretical assumptions that will allow the exploration of the normative EU climate diplomacy to take place.

2 Performance: The Choice of the Three-Levels of Analysis

The structural steps that have been proposed to analyze an entity's foreign performance are numerous. A common feature that these frameworks usually have is that they propose a three-level analysis to evaluate the level of performance, regardless of the factors they use to assess their findings (Baltag, 2018; Delreux, 2014; Jørgensen et al., 2011; Juncos, 2007; Oberthür & Groen, 2015). One of the most recent analytical approaches that explicitly emphasize the performance of the EU's external action is that of Blavoukos, proposed in 2015 in the book *The EU and the Non-Proliferation of Nuclear Weapons: Strategies, Policies, Actions*. Further, it was also applied and adequately appended both in a book edited by Blavoukos and Bourantonis in 2017, titled *The EU in UN Politics: Actors, Processes and Performances* and in a book edited by Blavoukos, Bourantonis, and Tsakonas in 2019 called *The External Relations of the EU*.

This model proposes the exploration of performance of the EU, rather than solely of its effectiveness, allowing a broader conceptualization by delving into 'the underlying, intra-organizational, agreement-reaching processes' (Blavoukos, 2015: 15). Moreover, it argues that the assessment of international performance provides a more comprehensive approach to how an international organization functions and achieves its goals, in addition to the actual policy outcome or output (Blavoukos & Bourantonis, 2017: 7).

Blavoukos' (2015) analytical framework to measure the EU's performance advances by examining three distinct levels, namely *output*, *outcome*, and *impact*. Output or micro-level is the result of the internal EU deliberations while formulating the EU's international position. Outcome or meso-level is the externalization of the results of the previous level. Impact or macro-level refers to 'the effects of the EU international outcomes' (Blavoukos, 2015: 21). All three levels are accompanied by several analytical factors, as well as parameters that guide the methodological direction of the framework (Blavoukos, 2015: 17).

Based on Blavoukos' work, an adjusted conceptualization of that scheme is proposed in order to be applied to the NPE discourse and offer a comprehensive framework to the normative identity of the EU throughout its performance regarding a particular subject. The three-level categorization remains but the focus diverges and concentrates on the *internal*, the *external*, and the *outcome evaluation*, terms that are differently operationalized than the previous ones, to comply with the needs of the NPE. 'Analytically, [the three-levels] constitute three distinctive steps in a causal chain of events' (Blavoukos, 2015: 17).

Recently, NPE theory has been intertwined with the broader discussion of performance and whether the EU's expectations are achievable (Larsen, 2020). In order to understand, unveil, and evaluate the expectations that the Union has when it negotiates climate targets with other international organizations (or with third actors in general), what has to be examined is the three distinct steps that will allow for a comprehensive understanding of what the EU as a whole intended to negotiate (expectations/internal processes), the means deployed to negotiate the agreed position (external processes), and the EU's ability to influence in favor of this position (outcome evaluation). Thus, *diplomatic performance* can be defined as a linear process that attempts to first explain the development of negotiating strategies, positions, and targets (intended or unintended); second, understand the activation of the diplomatic efforts; and third, move with the evaluation of the impact of the two previous steps. Through the above, we may comfortably support that performance simultaneously leads to enhanced (European) integration (Blavoukos & Bourantonis, 2013) and manifests a consistent external identity, the level of which usually depends on the achievement of the initial expectations and goals. In other words, when the original expectations are achieved, it is more likely to enhance the integrative aspects of the behavior of the EU Member States and to define the EU's (as a whole) external identity better.

Through this exercise, a contribution to the recent endeavors in bridging a gap that has been pointed out in the literature not long ago (Lavenex, 2014) is made. This gap is related to the connection that has been drawn between the internal policymaking on the one hand and the EU's external action and effectiveness on the other. The gap has also been observed in the literature of the EU as an international environmental negotiator (Delreux, 2009; Oberthür, 2011). Consequently, the analytical framework building on the three-level analysis that was just discussed proceeds by combining the internal policy procedures concurrently with the broader perspective of the structural aspects of the NPE (external action) and the discussion around their performance (evaluation).

3 Internal Aspect: Formulating Negotiation Strategies

In order to identify the strategies which emanate from climate norm(s) and construct the EU climate diplomacy, the assumption, made by other scholars who have written on the NPE, on internal processes among the Member States and the EU institutions, will be used (Bengtsson, 2015; Sjursen, 2006). This assumption postulates that internal processes are a necessary condition for the EU norms to be exported outside the EU. The internal aspect lies, again, on Manners' original conception of the EU's normative power (Manners, 2002: 240–241) and is further highlighted by De Zutter (2010), as evidenced in the previous chapter.

The interconnected relationship between the internal institutional structure and the external action that leads to influence is also recognized by Damro (2012) in his seminal article 'Market Power Europe' (MPE). Damro, building on Manners' NPE, attempts to construct a solid basis to explain how the EU can be perceived as an MPE, placing emphasis on the procedures of policy and decision-making as a result of the EU's regulatory character. He further supports that these procedures result in 'unintentional externalization, but when put in action, they clearly bolster MPE's intentional efforts at externalization' (Damro, 2012: 688). Accordingly, this assumption is applied by this book to NPE's function. The current Section 3 is dedicated to the operationalization of the internal processes that play a role in the EU's climate diplomacy when projected through the EU's normative complexion.

Competence, Coordination, Coherence: The Three COs

The difficulty that the EU usually faces when attempting to speak with a single voice in its international relations, is widely pinpointed in the literature (Juncos & Pomorska, 2020; Musu, 2003) including that of international climate negotiations (Delreux, 2014; Haug & Berkhout, 2010; Sjöstedt, 1998).

The most commonly used concept that aims to illustrate the Union's endeavor towards a single EU voice in its external relations is that of *coherence*, which has been characterised 'as more of an idea; a notion of unity in EU internal and external policy' (Lenk, 2015: 7). As such, coherence proves 'very difficult to measure' (Rasch, 2008: 205), as various preferences and divergences between the Member States have to be precisely analyzed. Moreover, there is literature supporting that coherence does not need to take into account the preferences of the Member States for the EU to act as a single actor in its external relations (Da Conceição-Heldt & Meunier, 2014a; Delreux, 2014), allowing for the focus to be on the ways common positions are decided through internal EU policy and decision-making processes.

Juncos argues that coherence needs *coordination* in order to be achieved. She then defines coordination as the function that would allow for an institution to 'work in an efficient and organised way' (Juncos, 2007: 23). Thus, before analyzing coherence, which is necessary in order to understand what the EU Member States wished to advocate at the international level and specifically to the two international organizations under examination, *coordination* needs to be examined. Coordination is much more tangible than coherence, as it illustrates the fact that the EU institutions and EU Member States are speaking with one voice in their external action, without the need to go in depth in regards to examining the preferences of the Member States, and without the requirement of the existence of common interests.[1] Besides, in the case studies under examination, the EU itself uses the word *coordination* to indicate the need for all its Member States to display unison in the IMO and the ICAO negotiations. Thus, the assumption is that there was a level of coordination in the respective case studies, as some of the promoted strategies were eventually adopted. Coordination, therefore, may precede but also

[1] On that observation, see Maurer and Raik (2018) on the role of the EU delegations in third countries.

succeed, coherence. It has to be mentioned that coordination has not been deeply analyzed as a subject in the field of EU studies, as it is the result of what usually happens behind closed doors in both formal and informal *fora*, creating a need to unveil the coordination process and its results, firstly by examining classified documents and secondly through the series of interviews with the involved EU Member States and EU institutions representatives.

Another reason that coordination is selected to be one of the main variables to be tested is the fact that numerous academic papers are delving into the coordination process of EU participation in the United Nations proceedings (in New York or Geneva), without, though, sufficiently theorizing it (Degrand Guillaud, 2009a, 2009b; Laatikainen & Degrand Guillaud, 2010; Smith, 2006). In addition, a smaller body of literature can also be found on UN Specialized Agencies, such as the International Labor Organization (Kissack, 2006, 2007). In order to extend the existing literature to include further analysis on EU coordination regarding the IMO and the ICAO, a focus on climate change policies will be made.

Furthermore, what is often missing from the existing analyses is the examination of another parameter, which is usually a reliable starting point for political scientists and not a factor for analysis. This factor is the EU's *competence* in a specific policy area. In the cases under examination, it is crucial to examine the competence of the EU institutions to govern climate change policies and strategies in the international transport sector, which—in itself—combines two different potential competence areas that are not yet exclusively transferred from the Member States to the EU institutions. As mentioned in the previous chapter, the controversies surrounding the climate policy competence field are significant. Therefore, the first part of the analysis will be devoted to examine *the three COs*, namely competence, coordination, and coherence. These factors shape the internal processes of the EU to construct a strategy and decide on how it will be subsequently externalized.

Before moving on with presenting the operationalization of the three COs, a reference to their relation with Europeanization needs to be made. It is often reported that the EU's external action has significant ramifications on the process of the EU's internal practices that lead to the deepening of cooperation between the EU Member States. Especially in climate change policies, alterations in the EU competences have been observed after significant international events (e.g., the Kyoto Protocol), which then, in their turn, influence the way the EU shapes its policies through coordination (Vogler, 2013). More interestingly, as Pomorska and Wright have noticed, 'socialization and learning [appear to be] potentially the most important mechanisms of Europeanization in foreign policy' with 'the coordination reflex and consensus-building acting as drivers of socialization' (Pomorska & Wright, 2013: 152). Therefore, the exploration of the three COs that strive for building a coherent and united EU position in climate diplomacy, substantially matches the discourse of Normative Power (Europe).

Competence

Two aspects of competence are examined: the EU competence on climate and environment issues and its competence to represent all Member States in the area of climate diplomacy.

Environmental (and Climate) EU Competence

Competence, both as a status as well as a political claim, is a subject that is inextricably linked with the European integration process (Garben, 2019). Competence is also a matter of constant move between the EU institutions and the EU Member States (Delreux, 2006), especially in climate change where the notion cannot be per se found in the Treaties, but is present in the EU policymaking processes and the policy implementation (Vedder, 2012). Specifically, in climate change which has gradually become a significant political priority of the Union, the 'competence creep' is of outmost importance in understanding how the EU institutions and the EU Member States cooperate in order to come up with a specific policy or negotiating position. 'Competence creep', as defined by Garben, is the process where 'the EU somehow manages to legislate and/or otherwise act in areas where it has not been conferred a specific competence' (Garben, 2019: 205). The discussion on the environmental (and climate change) competence has occupied both legal scholars and political scientists; while the former mainly focuses on the evolutionary course of the EU's competence, both the former and the latter focus on the ramifications of the competence that the EU has obtained in this policy area and to the extent of freedom that this competence provides to the EU institutions to act in the name of or instead of the EU Member States (Delreux, 2006; Doussis, 2019; Kulovesi & Cremona, 2014; Russo, 2017; Sjöstedt, 1998; Vedder, 2012).[2] The additional involvement of legal scholars with the political aspects of the environmental (climate) competence highlights the high politicization of the subject in a rather bold way and makes the process of exploring it more interesting.

Studying the process that the EU institutions follow aiming to acquire the competence to govern climate change policies and strategies towards the IMO and the ICAO is very important in the present context. First, it has to be established that the EU institutions do possess the competence to regulate climate change transport policies internally. This provides them with the opportunity to claim that they have the corresponding competence to regulate climate change transport policies externally as well. For all EU policies, this doctrinal function emerged back in 1971 during the CJEU seminal judgment, the so-called *ERTA Case* (Kulovesi & Cremona, 2014).[3] The ERTA Case 'justified the existence of external jurisdiction as a necessary consequence of either the corresponding internal jurisdiction or of the fulfillment of the Treaties objectives' (Doussis, 2019: 307). In other words, it 'created the doctrine of parallelism between internal and external competences' (Delreux, 2006: 234). The ERTA Case was a result of the so-called *implied powers* (Kadelbach, 2019), which is one of the main theoretical explanations of

[2] It has to be noted that the literature which focuses on EU external climate change competence is quite limited. The majority of the literature revolves around the EU external environmental competence by using climate change as example. As Vedder (2012) mentions, this might be happening because climate change policies, if implemented horizontally, would change the way the carbon-based system and everyday life works, thus creating an overly complex position within the EU system. It seems that he instigates to assume that EU Member States purportedly do not raise the issue of climate change competence, in order not to give the EU institutions access to their core state functions. Therefore, scholars do not have a lot to work with, concentrating on other case studies within the EU environmental governance.

[3] ERTA stands for European Road Transport Agreement 'on the work of crews on international transport vehicles' (for more information on the ERTA Case see, Kadelbach, 2019). For a brief analysis of the four cases that expanded the use of implied powers by the EU see Delreux (2006).

how international organizations operate (Blokker, 2009; Rama Montaldo, 1970).[4]

It is interesting to note that the ERTA Case concerned the transport sector. Delreux presents the disagreement between the European Commission and the European Council that ended up to the CJEU in a very comprehensive way:

> The CJEU was asked to rule on a dispute between the member states and the Commission about who should negotiate and sign an international agreement on road transport and safety rules for professional drivers. The Commission, arguing that the agreement would influence the internal transport policy of the EC, challenged the decision by the European Council that this agreement would fall under member states competences. The CJEU followed the Commission's claim and it states that when the EC has taken measures to realise common policy, the member states are no longer allowed to conclude agreements with third states that (could) undermine these internal EC measures. (Delreux, 2006: 234–235)

However, Delreux's interpretation of the extent of implied powers' application is more static and mentions that the CJEU is the body responsible for interpreting and determining the implied powers of the EU (Delreux, 2006: 234). Taking a more positivist stance, based on the theoretical basis of *functionalism* on how implied powers work (Imber, 1989; Klabbers, 2009), it can be argued that said implied powers which may be performed by the EU, emerge from the need of the Union to execute its functions, without needing the CJEU's validation; the CJEU is called upon to rule only if requested to do so.

The ability of the EU to vote as one actor or as an additional actor to the EU Member States during the negotiations that take place within the ICAO and the IMO, opens a further discussion on the EU competence puzzle.[5] To explore both the positive and the negative aspects of

[4] The implied powers of international organizations are not mentioned in their founding treaties, but without them, those that are explicitly stated could not be realized (for more, Klabbers, 2009: 59–64).

[5] The examination of the EU's competence to conclude international agreements has to be mentioned, as it is an extensively analyzed subject in the literature. These agreements fall under the status of 'mixed competence' or 'mixed agreements', meaning that both the EU Member States and the EU institutions have the competence to negotiate at the

such a possibility, the literature dealing with the conclusion of international agreements will be explored in order to shed light on how this might happen. It can be argued that the analysis of the processes of coordination and, especially, this of coherence that follows, would not have been so important if a clear competence structure was in place. Nevertheless, Member States could adapt to a new situation regarding the issue of exclusive EU competence in concluding climate agreements and find alternative ways to influence the process.

Russo, in an article titled 'Towards an Exclusive Competence of the EU to Conclude Climate Agreements?' argues that, 'from a strictly legal perspective' and 'from what scholars have considered concurrent competences', based on case law 'the EU competence to negotiate and conclude climate agreements could be considered as exclusive, and Member States would lose their external competences in the field' (Russo, 2017: 210). Russo recognizes that this, of course, is not the case yet and that the will of the EU Member States to keep this particular process under their control, would not allow formalization of the detachment of their competence in favor of the EU institutions.

Although the literature on the subject of the external EU environmental competence ostensibly adopts a more moderate position, it can be deduced to the same conclusion as that of Russo's (Buck, 2012; Doussis, 2019; Kulovesi & Cremona, 2014). EU Member States face several constraints, such as 'the duty of loyal cooperation' or the mandatory 'unity of international representation', and the direct mention of the climate change fight as an international EU duty, especially after the Treaty of Lisbon, was set in force. The disagreements, though, of several EU Member States in letting the EU exercise its full competence have led to a hybrid system where other factors, such as coordination and the achievement of coherence, play a significant role in the process of policy-making. Lastly, it is noteworthy to mention that the subject has not yet been brought to the CJEU, perpetuating a situation that does not make it easier for the Union to become a world leader in climate diplomacy (or, even, in other environmental issues).

international level and to sign the agreements concluded. This joint negotiating competence is often reported as the reason behind the moderate or even contradictory presence of the EU as a whole at the international climate/environmental bargains (Damro, 2006; Kulovesi & Cremona, 2014; Russo, 2017).

Chamon's position is that the Treaty of Lisbon recognizes the 'implied exclusive external competence' of the EU when the latter 'has exercised its competences internally' (Chamon, 2018: 1101–1102). Consequently, the first work assumption (WA) that emerges is the following:

WA1: The EU institutions have the competence to legislate for climate issues concerning the fields covered by the IMO and the ICAO and, therefore determine the negotiating position of the EU Member States and negotiate with both organizations for that purpose.

Acquiring Exclusive EU Competence

As pointed out in Chapter 2, combating climate change is classified as an environmental policy, following the provisions depicted in the Treaty of Lisbon. This fact has played a crucial role in the development of how the European system deals with combating climate change. Therefore, it is not surprising that academic literature engaging with the relevant issue uses the environmental EU competence as well as its external extensions as a starting point (cf. Doussis, 2019; Morgera, 2012). Having established that the EU has the internal competence to engage with climate issues *lato sensu*, the aim is to first examine if it can also extend this competence to the transport sector, and second to examine whether there is also an external extension that covers the EU–IMO and the EU–ICAO relations in this respect, that is approved and accepted by the Member States.

To explore the work assumption (WA1), we need to identify the factors that lead to the strengthening of the argument in favor of the exclusivity of the EU competence in the field of climate diplomacy. It has to be noted that this book accepts that the EU, in the context of the shared competences, does have such competence to negotiate agreements that aim to tackle climate change at the international level. The exploration of the matter comes after the observation of the refusal of the EU Member States to agree to this situation. Furthermore, during the initial interviews that were conducted while collecting material, confusion was identified in regard to the question of competence. Some officials argued that the EU has the competence to deal with climate targets in shipping, but there have been others who supported that the EU does not have the competence, even if they conceded that the Member States have to follow the common position agreed in Brussels. This ambiguity is also reflected in the literature (Gulbrandsen, 2013; Wouters et al., 2009), making its exploration necessary in order to understand the institutional underpinnings that govern the EU external climate transportation policy.

Recent literature on the competence puzzle has come up with several mechanisms through which the EU gains or enhances competences in an array of policy areas. The following Table 1 will highlight the work of Garben (2019) and Meunier (2017) on the EU competence creep.

In more detail, Garben (2019: 209–213) proposes several mechanisms that the EU can activate in order to obtain the desired competence, the most relevant for our case being this of *parallel integration* and this of *soft law*. Meunier (2017), coming from a different point of view, studies the procedure of how the nature of EU institution's competence for Foreign Direct Investments shifted and was included in the Treaty of Lisbon as exclusive. In her work, she looks more at the intentions and the preferences the EU actors have and the role that the latter play in configuring competence in a policy area. The aspects of two of the mechanisms that Meunier describes in her article that will be examined are those concerning *intergovernmental bargain* and *neo-functionalist spillovers* (split into *functional spillover* and *institutional spillover*).

Table 1 presents the aforementioned mechanisms (as well as the *expertise* of the Commission) in order to analyze whether the EU institutions have the competence to decide the negotiating position which the EU Member States promote in the ICAO and the IMO, externalizing the EU's climate norm (and, consequently, in other climate negotiations as well).

The problem of cooperation on climate change at the European level is diachronic, with scholars linking the competence of the European institutions with the coordination of the Member States to achieve coherence (Damro, 2006; Garben, 2019; Meunier, 2017; Oberthür & Roche Kelly, 2008). Although the competence issue, if resolved, would solve the dispute between the EU institutions and the EU Member States, as mentioned at the beginning of this section, EU competence is a dynamic situation, with each of the two competing parts perennially attempting to dominate over the other. Thus, since the EU competence is still a matter of concern, i.e., the ability of the EU to form the negotiating position in climate change policies towards the ICAO and the IMO is in flux, the process of coordination which attempts to amalgamate the diverging views of the Member States in the negotiating position that the EU institutions wish to perform, needs to be addressed.

Table 1 Mechanisms that explain the EU 'competence creep'

Parallel integration…	'… is constituted by action outside the EU framework, through intergovernmental processes governed by international law, undertaken by all Member States or a group of Member States with constitutional significance, when such international Treaty-making interacts with the EU legal/institutional/political framework. The action may complement, facilitate or support other EU policies, instruments and/or institutions and, as such, [be] intertwined with, and sometimes de facto entrenched in, EU action' (Garben, 2019: 212)
Soft law and soft political power	As one of the main goals is to give prominence to the European climate diplomacy, soft law has a significant role to play. Especially in the transport sector, where compliance with climate change measures is a relatively recent obligation, decision-making is largely based on internal procedures that do not meet hard law criteria. Moreover, as Terpan (2015: 70) showcases, soft law explains the hybrid model of the EU in which the relation between the political context and the EU legal order is often complicated and the dividing lines between them are not clear. Thus, apart from soft law that the EU institutions could possibly adopt, their soft power influences the positions of the Member States[6]
Intergovernmental bargaining	Through this process, which could also be connected to the coordination process that will follow in the next section, the EU Member States realize that their interests would be best represented by delegating the competence of negotiating international climate agreements (with other international organizations in our case) to the EU institutions
Functional spillover…	'… would place the causes of integration [of a certain policy area] in prior European integration in adjacent policy fields' (Meunier, 2017: 596)

(continued)

[6] This leads to the two next sections of this chapter, where competence can be obtained through the coordination that the EU institutions support in order to convince the Member States to speak with a single EU voice at the international level and through the coherence that can be a triggering factor to further the EU competence in a specific policy area.

Table 1 (continued)

Institutional spillover	'Supranational institutions put the competence shift on the agenda and engineer its passage' (Meunier, 2017: 597). This might as well happen through coordination. Both branches of the spillover effect match to the aforementioned implied powers of the EU. In the case that the EU institutions perceive that an issue will be better governed at the supranational level, and they already possess the competence to engage with neighboring policy areas, they will attempt to expand their scope to include the new topic
Expertise	A mechanism that can affirm the competence of the EU institutions to make decisions for a specific policy issue can emerge through their expertise (Vanhoonacker & Pomorska, 2013). Competence and expertise have been for long regarded as interconnected in climate change policies.[7] As tackling climate change has proven to be an overly technical problem (Doussis, 2017), the European Commission technocrats[8] have developed specialized groups to deal with it. In general, it can be argued that such groups have a stronger collective understanding of what is needed and what is feasible than any single national official. By establishing that its proposed solutions have the highest chances to prove successful in a negotiation, and—thus—be implemented at the next stage, the Commission enhances its 'competence creep'

Coordination

In the case studies under examination, the European Commission itself uses the word *coordination* to indicate both the need and the action taken for all the EU Member States to display unison in the ICAO and the IMO negotiations. Therefore, coordination is perceived as sequent to the competence of the Commission to deal with this issue. In our cases, it is *sine qua non* that, if the European Commission had an active exclusive competence on climate change policies in international transport, a move

[7] For an extensive study on the subject, see Dreger (2014).

[8] For the role of technocracy in exporting EU norms and rules globally, see Lavenex (2014).

towards coordination would not be necessary. However, since this is not the case, the Commission performs coordinating strategies to impound the Member States to the bargaining position that the former wishes to negotiate at the international level.

Using the concise definition given by Juncos (2007) as a starting point, namely that an institution needs coordination in order to 'work in an efficient and organised way', what will follow will present some of the characteristics mentioned in the literature which take place during the internal coordination in the EU processes. The first goal here is to examine the extent of top-down policy coordination led by the European Commission. The reason behind this decision is that coordination in the cases under examination, as well as during other international environmental and climate negotiations, does not only take place in official meetings in Brussels, but also in the corridors, behind closed doors, and on the spot at the headquarters of the international organizations under concern. More interestingly, the examination of the Commission's role is of paramount importance as the EU is not a full member of neither the ICAO nor the IMO and, therefore, its interests may only be represented by and are depended on what the EU Member States decide to advocate on at the stages of the two organizations.[9]

Methodologically speaking, the extensive work of Kissack (2006, 2007) on coordination pinpoints that it is exceedingly difficult to have a solid view of the coordination processes that take place within the EU, because of all the aforementioned reasons and the usual lack of official documents portraying the discussions held during these *informal* meetings. Thus, the focus regarding coordination, as stated earlier as well, will be on what happened after the adoption of the CORSIA and the Initial IMO Strategy, which took place quite recently and which are seminal in the course of climate change protection in the two IOs.

In this part, two distinct work assumptions emerge. The first refers to the effect that internal coordination has on the external performance of the EU as a whole; the second refers to the effect the same process has to the internal EU governance.

[9] Giving attention 'to the role of the Commission in policy coordination [will allow us] to fully capture European integration and governance dynamics' (Mattocks, 2018: 320).

WA2: Internal coordination is needed in order to reach an agreement on the negotiating position and strategy.
WA3: Internal coordination is necessary to strengthen EU climate competence and the European Commission performs it to strengthen its position in the EU climate governance structure.

The recent article of Mattocks (2018), which deals with the relation between 'the European Commission and policy coordination' provides a starting point to examine the aforementioned work assumptions. Mattocks notes that this kind of coordination, as well as its consequences, have not been studied adequately.[10] Mattocks studies policy coordination within the intergovernmental process of the Open Method of Coordination,[11] in a field where the EU only has a supporting role. In this book, the focus is on policy coordination in a field where the EU competence is shared (or, even, exclusive as some scholars suggest) and in a field that it is not only governed by intergovernmental processes, but also by supranational institutions.

Mattocks (2018) argues that coordination is rather a political act in contrast to the administrative aspect that was supposed to vest it when initially presented in the *White Paper on EU Governance* in 2001.[12] She further argues that the process of coordination allows the Commission to decide on several issues, ramming that way the prevailing intergovernmental core of the European governance. While the majority of the literature points to the fact that the Commission has seen its power decreasing in regard to the creation of the agenda (cf. Kreppel & Oztas, 2017; Ponzano et al., 2012), the climate field ought to be addressed from a different perspective, since all the facts presented in the previous chapter lead to the conclusion that the Commission moves towards the strengthening of its competence in the field.

As the coordination procedures are pivotal to forming common negotiating positions, two methods identified by Schout and Jordan (2005) on

[10] Apart from mentioning it in the article (Garben, 2019: 318), this is also proven by the extensive literature that she uses in which just a few references dealing with the issue are included (Garben, 2019: 330–334).

[11] The open method of coordination 'is a form of intergovernmental policymaking that does not result in binding EU legislative measures and it does not require EU countries to introduce or amend their laws' (EUR-Lex, 2020).

[12] For the White Paper, see Official Journal of the European Commission (2001).

how the Commission coordinates the EU Member States will be applied. The first is that of *hierarchy*, that is the adoption of European legislation that constrains the Member States and the second is *networks*, that is, the networks that are developed within the EU governance process in which both European and national representatives participate.[13] An additional method regularly mentioned in the literature of EU coordination, this of *softer mechanisms*, such as 'day-to-day management' (Mattocks, 2018) and *persuasion*[14] will be examined throughout the analysis. These mechanisms, as well as the aforementioned adoption of legislation (which will be examined according to what was discussed in the section regarding competence), intertwine perfectly with the mechanisms of the NPE approach.

Coming back to the concept of networks,[15] the deliberations between the European Commission and the EU Member States could be analyzed in a deeper level and shed light to the fine details of the bargaining procedures that lead to possible common negotiating positions. Recent literature links this with negotiations theory and the agenda setting literature (Dikaios & Tsagkroni, 2021), naturally allowing to examine the internal EU (climate) diplomacy in the cases under examination. The four-step path proposed by Dikaios and Tsagkroni (2021) allows to examine not only how the Member States managed to reach an agreement on a common negotiating position towards the IMO and the ICAO, but also to examine the efforts made by the EU institutions and the Member States themselves to convince one another to reach this agreement.[16]

[13] Schout and Jordan mention another coordination mechanism as well: market(s). This one is out of the scope of this book. Knill and Tosun (2009) apply the same mechanisms in order to explain 'how does the EU shape environmental policy adoptions within and beyond its borders'.

[14] Several scholars classify 'persuasion' as a soft mechanism in EU external affairs (Börzel & Risse, 2012; Botta, 2014; Garcia & Masselot, 2015).

[15] For example, see the reference list of Schout and Jordan (2005), or the works on Actor–Network Theory that may offer a new path to how the EU Member States and the EU institutions can influence each other (e.g. Callon, 1984; Dikaios & Tsagkroni, 2021). On the difference between hierarchical modes of policymaking and policymaking through networks, see Ingold (2011).

[16] Here, as an additional explanation, the extensive literature on the socialization processes that take place within the European structure for the national representatives could be mentioned, as well as the European officials (cf. Juncos & Pomorska, 2011; Lewis, 2005).

The goal through the coordination process is to understand what happens at the intergovernmental level when the supranational attempts to influence the former and whether there is any peer pressure among the Member States. Vanhoonacker and Pomorska (2013) suggest a four-step categorization of how actors in the European foreign policy field gain attention and build credibility to establish themselves as preponderant to deal with an issue. Dikaios and Tsagkroni (2021), building partly on the said categorization, explore the way actors build and use the adequate network to pursue their goals. As the European Commission attempts to prevail upon the Member States and be the one who decides what is negotiated at the ICAO and the IMO, the following combination and interpretation of the above analytical frameworks will be applied.

Through Table 2 (see next page), the ability of the EU as a whole to negotiate at the international level, as well as to stipulate the negotiating agenda and strategy is highlighted. This, of course, gives prominence to the role of the European Commission in the process of coordination, since not being a full member of either ICAO or IMO would not allow it to perform its line of bargaining in a more self-prominent role. As this is not the case yet, in the next subsection the focus will be on whether the coherent appearance of the EU Member States *vis-à-vis* these two organizations has a significant role to play in externalizing the EU's positions.

Coherence

As a concept, coherence might be a hard road to go down exploring, as it seems to be more of an ideational feature of the EU's united appearance (Lenk, 2015; Rasch, 2008) and no agreement between scholars exists on its conceptualization (Thomas, 2012). Nevertheless, coherence or speaking with a single EU voice in the case studies under examination is of outmost importance, and, as such, the emphasis is placed on the measurable *single EU voice* (founded on the internal processes of the EU) that emerges through the functions of competence and coordination.

The reason coherence is included in the internal processes, although being a de facto external action, is that it portrays an important unity among the EU Member States and the institutions. However, speaking with a single voice has significant ramifications for the European integration and the EU policymaking (Gebhard, 2011), making it that way inextricable part of the internal processes conceptualization.

Table 2 Mechanisms that explain the EU 'internal coordination creep'

Setting the agenda and framing the problem(s) and the roles	At this first stage, the European Commission declares its intention to deal with a specific policy issue. It then prepares the position it wishes the EU as a whole to negotiate externally and requests the EU Member States to agree on it and to bind themselves to the joint diplomatic efforts. According to Vanhoonacker and Pomorska (2013) this can be done either by using 'big words' or by a 'small-step approach'. The first clearly and outrightly sets a specific target and the means for this target to be achieved, as well as the greater framework within which the issue will be dealt. The second gradually builds the position at stake by forwarding droplets of the subject under concern in various venues. Both these strategies have the ultimate goal of 'imposing' the Commission's position on the Member States and the Member States deferring to the Commission's guidelines on how to act

(continued)

Table 2 (continued)

Internal negotiations	Both strategies in the previous cell of the Table usually lead to long-lasting discussions among the EU Member States and the EU institutions to adopt, reject, or alter the position proposed by the Commission. As no agreement on the EU climate (transport) competence has been reached, this intergovernmental stage is necessary and is of utmost importance for the negotiating position to be agreed. At this stage, the efficiency and will of specific actors to deal with the subject presents itself. In addition, who is going to be representing the group is decided.[17] Moreover, the final negotiating position is adopted, and, depending on the issue and the wording of the decision, Member States and/or EU institutions have to promote and speak for it externally
Establishing jurisdiction	Through these activities the Commission facilitates not only the coordination among the EU Member States, but also the transference of certain powers to itself, aiming at some point to secure its exclusive competence in negotiating international environmental (and climate) agreements

[17] In the cases under examination, this is already known as the EU as a whole is not a full member to either of the two IOs. Nevertheless, it is interesting to examine the dynamics that emerge not only between the Commission and the Member States but also among the EU Member States.

As put by Thomas, the concept of coherence is 'able to accommodate empirical variation, [...] generalizable across issue areas and reasonably parsimonious given the organizational complexity of the EU' (Thomas, 2012: 459). Thomas goes one step further and operationalizes coherence by proposing two ways of exploring it; first, through *policy determinacy* and, second, through *political cohesion* of the subject under concern.

In our case, as the goal is to explore the establishment of the negotiating position of the EU as a whole, this exploration will take the following form: *negotiating determinacy* and *diplomatic cohesion*. On the one hand, at the internal level, negotiating determinacy refers to the explicit structure of the negotiating position that confines the Member States to support a certain line. On the other hand, at the external level, *diplomatic cohesion* refers to the extent to which EU Member States and institutions advocate the common negotiating position that has been concluded through the previously presented processes. Sequentially, voting cohesion plays an important/decisive role as well.

Apart from those two factors, there is a third factor which has recently appeared in the literature named *coheritization*. According to Schmitz and Eimer (2020), coheritization is the action of the European Commission to vocally support certain positions that it wishes to promote and establish. Coheritization, therefore, seems to entail two functions for the cases at hand. First, the previously presented actions, namely all those things that the Commission does to gain more competence in combination with coordination, eventually lead to a single voice. Second, it implies that through a coherent EU action in a particular policy field, the Commission gains an additional political competence from the Member States and is able to insist on respective coherent action in the same or related policy fields. This latter function concurrently enhances the integration among the Member States and the EU institutions in this field.

According to Thomas, the barrier of appearing coherent has been lifted since the Treaty of Lisbon was put into force. Thomas cites Article 24(3) that explicitly highlights that 'the Member States shall support the Union's external [...] policy actively and unreservedly in a spirit of loyalty and mutual solidarity and shall comply with the Union's actions in the area'. Therefore, the goal here is to examine whether the Member States of the EU speak with a single voice at the ICAO and the IMO, as this was decided during the internal coordination process. As put by Blavoukos et al. (2016: 2) 'coherence is conceptualised as the ability to reach a common position and to present it with a single voice'. The second aim is to examine the extent to which the single EU voice leads to coheritization diachronically. In other words, to examine whether previously coherent positions towards the IMO and the ICAO regarding climate change lead to more coherent appearances in the future and whether this action enhances the role of the European Commission in this particular framework.

Thomas notes that 'coherence may be necessary for the EU to exert its influence abroad' (Thomas, 2012: 472), but is also necessary for the EU institutions to appear as a whole, as a united and unified actor. A volume of literature intertwines the notion of coherence with the external effectiveness of the Union and, although some critics claim that coherence is not a decisive factor in explaining the EU's external effectiveness as mentioned earlier, coherence is perceived as a crucial component of the Union's action. Lastly, it is important to say that for scholars writing on coordination, coherence (*or cohesion or speaking with a single voice*) is the result of the internal EU coordination processes (Degrand Guillaud, 2009a; Kissack, 2006, 2007; Panke, 2017; Smith, 2006).

When emphasizing the competence status and the coordination process as to how the EU Member States end up supporting a common negotiating position, the appearance of coherency is almost as important as the success of the coordination process. Therefore, the work assumption to be tested is:

> *WA4: Coherence is based both on the status of the EU competence on the subject and on the internal coordination results.*

Fig. 1 shows this connection.

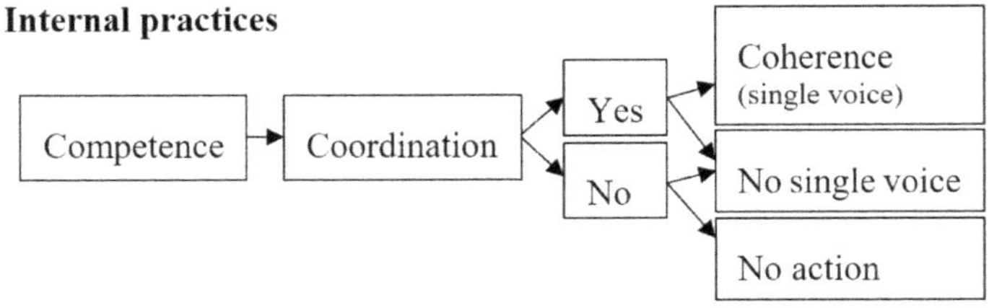

Fig. 1 Proposed composite model for the EU internal practices

European Cooperation: The Three COs

Based on the analysis above, it becomes evident that *the three COs* create the necessary preconditions for the EU as a whole to cooperate and perform as a normative power at the global level.

*Co*mpetence is of particular interest, as the complexity that the institutional framework of the issue under consideration is high. The European Commission has the shared competence to develop climate policies both internally and externally, as well as to negotiate and sign international environmental (and climate) agreements. On the other hand, the situation in international organizations is different, as there are no agreements to be signed, but the decisions are taken within the institutional bodies of the organizations. Thus, the Commission, which has specific targets regarding mitigating GHG emissions from the global transport sector, is dependent on the EU Member States to negotiate as members of those organizations the common position that has been agreed upon through the internal EU coordination.

*Co*ordination is the least examined aspect of the three COs,[18] something which is expected, since coordination often takes place behind closed doors and on the spot, and it is usually informal, meaning that no documents are resulting from its processes. In order to fully understand and give prominence to this process, it is essential to firstly understand the extent of the role it plays and, secondly, to establish it in the theoretical discussion of the EU internal processes in general. The relevant analysis will be based mainly on the evaluation of the information gained by interviews with the respective stakeholders.

Moreover, the internal coordination process suits the definition that Keukeleire gives to the internal diplomacy of the EU that is, 'the settlement of mutual relation among Member States' (Keukeleire, 2003: 32). Thus, the examination of the internal EU's attempts in externalizing climate norms to the ICAO and the IMO can be characterized as *internal EU climate diplomacy* towards the said organizations. As internal coordination cannot stand alone within the European governance process, competence gives the Commission the necessary basis to put forward

[18] Articles that engage with the internal processes of the EU in the field of environmental and climate external EU action, mostly focus on the aspects of competence and coherence (see, cf. Delreux, 2006, 2014; Van Schaik, 2010) or by dealing with the internal without analyzing it (Da Conceição-Heldt & Meunier, 2014a).

a particular strategy and coherence depicts the result of the internal coordination process (Gebhard, 2011).

The literature on *Co*herence makes regular references to internal processes that take place in order for the EU to form the discussed single voice, without, though, giving emphasis to these processes themselves, with only a few scholars attempting to interrelate them (Kissack, 2006, 2007). This book will attempt to fill in that gap by paying attention to internal coordination processes prior to and during the negotiations under examination. This way, speaking with a single voice will rely on new theoretical foundations (a combination of competence and coordination as conceptualizd above), pointing out the long institutional process to reach an agreement.

Through the analysis of this part, we can confidently support that the three processes, the three COs that take place within the institutional European framework, build the conditions for the EU to exert its influence globally on a specific subject through a general (normative, as established earlier) direction. These processes are mostly driven by the EU institutions that wish to be in control of the European affairs and to appear in the international arena as a single actor.

4 Normative Power Europe: A Framework of Externalizing Norms

As it has been demonstrated in Chapter 2, a part of the theoretical approach that will be applied to identify the action that the EU performs in order to externalize its policies will be this of the NPE, grafted with the constructive aspects of climate diplomacy, as described earlier.

In the previous chapter, what was also highlighted was the significant impact that NPE had on European Studies, as well as to the discussion on the identity of the EU. NPE, as conceptualized by Manners in his pivotal work 'Normative Power Europe: A Contradiction in Terms?' (2002), apart from setting the foundations for a new perspective on the kind of power the EU displays in world politics, it also attempts to create and establish a greater narrative of how the EU should act in its external relations. As also showcased previously, scholars attempting to move forward this discussion propose new mechanisms that might explain the EU's normative power, without, however, consulting Manners' mechanisms or even devising their own. Nevertheless, the main reason behind the application of Manners' mechanisms in their original operationalization is that they offer a comprehensive approach to explain how the Union acts in its international affairs when it wishes to pursue a certain goal by projecting a norm. A second reason is that the majority of other mechanisms as proposed in the literature, which aim to explain the EU's action as a normative power (or broadly in its external affairs) may fall under the broader operationalization that Manners initially suggested (see Chapter 2). The third reason is that other conceptualizations of how the EU promotes its rules contain coercion (or similar concepts) (Damro, 2012; Lavenex, 2014), whereas in our case studies, this is not applicable, since international organizations cannot be forced—through sanctions for instance—to adopt any particular policy. The mechanisms of NPE seem to better match the EU's behavior towards international organizations, especially on a subject such as climate change which has been set a priority for the EU in the last years and its combating has *par excellence* normative and regulatory implications.

Thus, the main theoretical challenge is in what way the NPE and climate diplomacy will be defined and operationalized.

Manners (2002) set the foundations for a new perspective on the kind of power the EU displays in world politics. The main argument of this approach is that the EU 'can be *conceptualized* as a changer of norms in the international system, [it] *acts* to change norms in the international system, and [it] *should* act to extend its norms into the international system' (Manners, 2002: 252).

In this book, this argument will be addressed partly, as the answer of the normative question 'should the EU act to extend its norms into the international system?' does not fall into its scope. Instead, the strategies emanating from the norms that the EU attempts to project and to promote in the international system will be explored. To that end, in this part, attention will be paid to the way the Union acts in its endeavor to be a norm changer at the international level. Manners indicates six factors that explain how the EU norms are being diffused. These six factors are presented in Table 3 (see next page).

In order to draw the most useful conclusions, Manners' factors are going to be detached from their conceptual ideal, which postulates that they explain the ways the Union acts in its endeavor to become a certain type of power player in the international arena. They will be instead perceived as mere mechanisms that explain the EU's function within a specific framework of activity addressed to third parties. In other words,

Table 3 Factors that explain the EU norm diffusion (according to Manners)

Contagion	'Diffusion of norms results from the unintentional diffusion of ideas from the EU to other political actors'
Informational diffusion	'... is the result of the range of strategic communications, such as new policy initiatives by the EU, and declaratory communications, such as initiatives [from EU institutions]'
Procedural diffusion	'... involves the institutionalization of a relationship between the EU and a third party, such as an inter-regional co-operation agreement, membership of an international organization or enlargement of the EU itself'
Transference	'Diffusion takes place when the EU exchanges goods, trade, aid or technical assistance with third parties through largely substantive or financial means'
Overt diffusion	'... occurs as a result of the physical presence of the EU in third states and international organizations'
Cultural filter	'... affects the impact of international norms and political learning in third states and organizations leading to learning, adaptation or rejection of norms'—"is based on the interplay between the construction of knowledge and the creation of social and political identity by the subjects of norm diffusion'

Source Manners (2002: 244–245)

this book does not attire these factors with the paramount goal of creating an ideal type or a grand explanation of how the EU operates, but it deals with them as mechanisms of EU's external action within a wider framework. This argument is supported by the work of Lavenex (2014) who writes on how the EU exports its rules outside its borders. In her work, while perceiving norms as a powerful way to influence third actors, she develops slightly different mechanisms to explain how the EU externalizes its norms, but to a large degree, she uses and extends Manners' operationalizations. For example, Lavenex applies *learning and socialization* as a mechanism to explain 'how EU rules travel' that coincides with Manners' cultural filter or *emulation* that, broadly perceived, can fall under the *informational diffusion*, as the analysis will showcase.

As the core case of the book is the relation between the EU and the ICAO and the EU and the IMO, the research will examine if the six factors are present in the EU–ICAO and the EU–IMO relations. It has to be noted that, when Manners applied this model to his case study (abolition of death penalty), only three of these factors were confirmed (Manners, 2002: 252). This did not discourage him to support that the EU is a normative power; rather it made him make the case that this model is not the sole way to analyze the case. Yet, because the nature of the EU climate diplomacy is still under formulation, the goal is to demonstrate that it can be seen as an integral part of the broader conceptualization of the NPE.

Lastly, throughout the analysis on the externalization of the EU climate norm towards the two international organizations, and in parallel with the examination of the above factors by Manners, ***diplomatic cohesion*** will also be tested.

5 Diplomacy Evaluation

In the previous sections, the way a norm is externalized and the factors that make this externalization a reality were unfolded. This section will complete the analytical framework and will allow us to examine whether the EU can perform a normative climate diplomacy, complementing the two previous analytical stages. The effectiveness of the EU in externalizing climate norms and rules to the ICAO and the IMO will be examined, giving emphasis on the role coherence played. Additionally, light will be shed on whether the result is more effective when the EU appears as a coherent actor in the international arenas, by concurrently measuring the effectiveness of the EU to externalize climate norms through the six NPE mechanisms as a whole and separately as Member States. Subsequently, focus will be given on whether the EU performs a normative climate diplomacy through the analysis of its actions, which aim to result in influencing the 'normality' of other actors.

Effectiveness of the External EU Action

The concept of effectiveness has occupied a significant part of the literature in EU foreign relations and external action, especially during the previous decade and specifically within other international institutions. Romanyshyn's (2015) article, titled 'Explaining EU Effectiveness in Multilateral Institutions', offers a comprehensive literature review on the subject, combining the existing knowledge so far on the overall performance. While Romanyshyn does not offer a new definition of effectiveness, pointing out that 'measuring [it] is a quite difficult exercise' (Romanyshyn, 2015: 877), his article allows and justifies for the classical definition as it has prevailed in this literature.

Thus, effectiveness is defined as the achievement of the originally posed goals[19]; in negotiations, that is, to convince your interlocutors to accept and adopt your positions and targets. This definition is similar to what Thomas suggests, i.e., 'the Union's ability to shape world affairs in accordance with the objective it adopts on particular issues' (Thomas, 2012: 460) and to what Oberthür and Groen propose, i.e., 'goal achievement refers to the extent to which EU objectives are reflected in the international outcome, irrespective of actual EU influence' (Oberthür & Groen, 2015: 1324). Both definitions have significant normative origins—although they are not mentioned—that can be equally extended to the definition applied here. After all, the core of normative power lies in the modification of what is considered at a given moment to be 'normal', whether the influence was direct or indirect.

As the internal practices that lead to coherence are examined, it is inevitable that the first part of measuring the effectiveness of the EU's action in the IMO and in the ICAO will begin from the significant part of the literature that reflects the connection between two concepts—coherence and effectiveness. This literature stretches to two directions: the first examines the relation between the two concepts and the second analyzes whether their link is sufficient to explain the role that the single voice has in effective externalization of EU positions (Da Conceição-Heldt & Meunier, 2014b; Smith, 2006). The question that derives naturally from these observations is to what extent does the single EU voice lead to results closer to the Union's initial negotiating positions. In other words, to what extent is the EU effective, when acting coherently?

While the literature argues extensively that the external effectiveness of the EU is not adequately explained by the fact that the Union appeared coherent at the international stage (Da Conceição-Heldt & Meunier, 2014b; Macaj & Nicolaïdis, 2014; Panke, 2014; Thomas, 2012), in international environmental negotiations it seems that coherence has indeed had an effect on whether the EU will realize its objectives (Delreux, 2014). In order to clarify whether the EU was effective in its endeavors,

[19] The literature that defines effectiveness in EU external relations and especially in its performance within other international institutions through goal achievement is voluminous. Some of this literature is the following: Delreux (2014), Jørgensen et al. (2011), Oberthür and Groen (2015, 2018), Romanyshyn (2015).

a comparison of them with the final outcome (or the final output) at the respective international organizations needs to be drawn (Blavoukos, 2015; Da Conceição-Heldt & Meunier, 2014a; Oberthür & Groen, 2015). The work assumption to be analyzed is the following:

WA5: The EU is externally effective when it speaks with a single voice.

The correlation between coherence and effectiveness serves to explain the results that the internal EU arrangements allow for an effective external action. The effectiveness, though, in our cases is also reliant on the sufficient influence that the mechanisms that externalize norms have on other actors. Therefore, the effectiveness of the external practices that the EU performs within the framework of NPE, will be examined in a case-by-case manner, meaning that every mechanism—at any given point—that appertains to the EU's diplomatic efforts, will have a reference to its results. This adds to the definition of effectiveness given just above, i.e., the achievement of the originally posed goals. Therefore, a second question to be answered, emerging from this section, is that of the extent to which the NPE mechanisms achieve their objectives. The subsequent work assumption is:

WA6: The EU is externally effective when the NPE mechanisms apply in the case under consideration.

Thus, effectiveness is measured by the successful externalization of the EU climate norm on the basis of the criteria/mechanisms that have been presented in Section 4 of this chapter, on the one hand, and by the coherent appearance of the EU Member States on the basis of the functions unfolded on section 2 of this chapter, on the other. What is interesting to note is that, although up to that point, the role of the European Commission seems to be prevalent, the burden to advocate a single position in the ICAO and the IMO falls upon the Member States, whereas the burden for a successful externalization of a norm is shared.

Lastly, a factor that has to be mentioned is that of the international or external context within which the EU acts, which might have a severe impact on EU's external effectiveness. This factor has been increasingly popping up in the literature (Larsen, 2014; Romanyshyn, 2015; Torney, 2014) but it has not yet been an issue of major focus. This might have two explanations. First, it can be exceedingly difficult to study the set of *external perceptions* that others have for the EU in a particular context regarding a specific subject; even more so when this focus is on international institutions where the number of the Member States is high. Second, the majority of the analytical frameworks developed to explain the EU's function focus on the actions that the EU performs to achieve an objective. Although interesting, the first explanation falls out of the scope of the research at hand, since it is not relevant to the goal of the book which is to explain how the EU develops a negotiating position, how it externalizes that position and whether such position has had an impact on the outcome (of the negotiations with the two organizations under concern).

Having suggested all the above, Fig. 2 illustrates the analytical framework in a straightforward manner.

Finally, as the broader goal of this part is to analyze the effectiveness of the EU as a normative power in the field of climate diplomacy, what has to be mentioned (and will be the underlying logic during the analysis), is that norm diffusion is a constant procedure that often has no instant results or, even, the expected results at all. Therefore, one cannot and must not measure the research object's effectiveness immediately, when the study has to do with norms, but to reasonably expand the time frame in order to understand and create a more comprehensive picture of a particular function. This exercise, while not popular with politicians and the wider public, who usually want immediate deliverables and tangible change,[20] for scholars it can offer significant tools to understand and analyze the

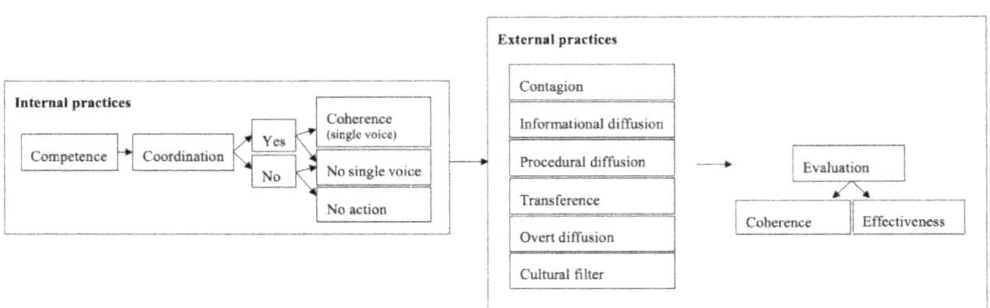

Fig. 2 Proposed composite model for EU normative climate diplomacy

[20] Also noted in the international climate politics (Oberthür, 2011).

timely process that norm diffusion takes. Consequently, apart from the analysis of the EU's external performance based on the aforementioned factor of effectiveness, the storyline of the bargaining procedures and the development of the EU international climate policy towards the ICAO and the IMO will showcase the extent of the EU's influence to these two organizations.

6 NPE and EU Climate Diplomacy

The relation between the EU and the ICAO and the EU and the IMO in the selected case studies can be seen as part of the broader EU climate diplomacy. As it was demonstrated previously, the traditional notion of diplomacy is under scrutiny. Hence, climate diplomacy may be interpreted as a distinct field within foreign affairs. This is also consistent with one of the initial assumptions of the NPE, which maintains that a normative power is 'characterised by common principles and a willingness to disregard Westphalian conventions' (Manners, 2002: 239). Common principles may count as an initial thought on internal coordination, while the contempt of the Westphalian system, which concentrates only on hard politics, is achieved through the diffusion of norms. Moreover, the Westphalian approach is also challenged by the tightly intertwined relationship between the last factor of the redefinition of diplomacy (augmentation of actors) and the three factors out of six of norm diffusion (procedural diffusion, overt diffusion, cultural filter). This is because the latter refer directly to relations between the EU and other international organizations and institutions, a fact that would be absurd in traditional diplomatic negotiations where sovereign states are the only actors.

Consequently, having suggested that the EU climate diplomacy fits the NPE theoretical discourse, and having demonstrated that in order for NPE to display tangible results there has to be internal coordination and the fulfillment of at least some of the six factors of Manners, the following two umbrella work assumptions have to be assessed and proven in order to demonstrate whether the EU functioned as a normative power in externalizing the EU climate norm towards the ICAO and the IMO:

- *EU climate diplomacy is grounded on internal political processes.*
- *EU climate diplomacy diffuses EU climate norms.*

Table 4 Work assumptions

A: EU climate diplomacy is grounded on internal political processes
WA1: The EU institutions have the competence to legislate for climate issues concerning the fields covered by the IMO and the ICAO and, therefore determine the negotiating position of the EU Member States and negotiate with both organizations for that purpose
WA2: Internal coordination is needed in order to reach an agreement on the negotiating position and strategy
WA3: Internal coordination is necessary to strengthen EU climate competence and the European Commission performs it to strengthen its position in the EU climate governance structure
WA4: Coherence is based both on the status of the EU competence on the subject and on the internal coordination results
B: EU climate diplomacy diffuses EU climate norms
WA5: The EU is externally effective when it speaks with a single voice
WA6: The EU is externally effective when the NPE mechanisms apply in the case under consideration

The reason that these two propositions are characterized as umbrella work assumptions is that the work assumptions established so far throughout this chapter can be grouped under these two, formulating a work assumptions list (Table 4).

7 Conclusion

This chapter has outlined the theoretical framework on the basis of which the empirical data of the book will be analyzed by introducing relevant alternative concepts to the already established concept of performance, in the endeavor to showcase the role that NPE plays in the latter.

As illustrated, performance is understood as a process that allows drawing a comprehensive image of the EU's diplomatic function. Naturally, the proliferation of the studies focusing on the EU's external relations, affairs, and policy has led to more fragmented analyses that focus on distinct sectoral subjects.[21] This might also be a result of the nature of the EU, whose policies have sprawled towards all directions, regulating the majority of issues in politics. As the EU is conceived often as an international organization whose practice has external ramifications

[21] For example, see the volume edited by Damro et al. (2018) on the EU's sectoral external engagement.

that extend further than just to its Member States, it is essential to study the influence it projects internationally. Thus, a three-level analysis for performance is adopted, which will produce a complete framework that will strive to explain where this influence comes from, how it is projected, and whether it can indeed have an impact.

The analytical framework is in dialogue with previously developed frameworks that give emphasis to different concepts, mechanisms, or factors of analysis. The proposed framework aims to add a new view to this discussion by exploring EU relations between the IMO and the ICAO on the subject of developing climate change strategies by the two organizations. The focus is on the aspects of internal and external EU climate diplomacy in the international transport sector and the concepts selected in this chapter encourage the further exploration of these particular relations. This study of the internal and external EU climate diplomacy will also allow us to answer the fundamental research question, i.e., in what ways and to what extent does the EU perform a normative climate diplomacy. The holistic approach that this framework aspires to offer additionally refers to the conceptualization of the NPE as a comprehensive approach. In order to grasp the core elements of its influence, the approach needs to be injected with aspects of the internal EU policy-making processes, adding this way to the scholarly discussion of whether normative EU power is still relevant and what are the ways to keep it up to date.

The chapter begun by examining the legal and political basis of the ability of the EU to come up with strategies aiming to influence the greenhouse gas emissions mitigation governance globally. It then moved to internal EU coordination, as the primary process which defines the final decision that the EU will promote internationally; internal coordination can be perceived as the core element of the EU internal climate diplomacy process. Further, the frequently contested concept of coherence is analyzed as, in this case, it plays a significant role, since the EU acting as a whole provides a significant benefit to the prospect of the desired outcome by negotiating as a coherent actor within a multilateral system. The NPE mechanisms that explain the externalization of a norm, which include both direct and indirect diplomatic methods of influence, were also shown to depict adequately the EU external climate diplomacy. Finally, the results of speaking (or not speaking) with a single EU voice are examined. The chapter is concluded with a discussion on how the analytical framework interrelates with the concept of (EU) climate diplomacy, concurrently grouping the main propositions of the book. The following chapters will empirically examine the application of the analytical framework that has been employed, as well as the work assumptions.

References

Baltag, D. (2018). EU External Representation Post-Lisbon: The Performance of EU Diplomacy in Belarus, Moldova and Ukraine. *The Hague Journal of Diplomacy, 13*(1), 75–96.

Bengtsson, R. (2015). The EU and Global Economic Governance: Playing the Role of a Global Leader? In A. Bjorkdahl et al. (Eds.), *Importing EU Norms: Conceptual Framework and Empirical Findings* (pp. 39–54). Springer.

Blavoukos, S. (2015). Capturing the EU's International Performance: An Analytical Framework. In S. Blavoukos, D. Bourantonis, & C. Portela (Eds.), *The EU and the Non-Proliferation of Nuclear Weapons: Strategies, Policies, Actions* (pp. 12–30). Palgrave Macmillan.

Blavoukos, S., & Bourantonis, D. (2013, May 9–11). Struggling with Performance: An Analytical Framework for the EU International Interactions. *Paper prepared for the EUSA Thirteenth Biennial Conference.*

Blavoukos, S., & Bourantonis, D. (Eds.). (2017). *The EU in UN Politics: Actors, Processes and Performances*. Palgrave Macmillan.

Blavoukos, S., Bourantonis, D., Galariotis, I., & Gianniou, M. (2016). The European Union's Visibility and Coherence at the United Nations General Assembly. *Global Affairs, 2*(1), 35–45.

Blokker, N. M. (2009). International Organization or Institutions, Implied Powers. In R. Wolfrum (Ed.), *Max Planck Encyclopedia of Public International Law*. Oxford University Press.

Börzel, T. A., & Risse, T. (2012). When Europeanisation Meets Diffusion: Exploring New Territory. *West European Politics, 35*(1), 192–207.

Botta, M. (2014). Competition Policy: The EU and Global Networks. In G. Falkner & P. Müller (Eds.), *EU Policies in A Global Perspective: Shaping or Taking International Regimes?* (pp. 76–92). Routledge.

Buck, M. (2012). The EU's Representation in Multilateral Environmental Negotiations after Lisbon. In E. Morgera (Ed.), *The External Environmental Policy of the European Union: EU and International Law Perspectives* (pp. 76–95). Cambridge University Press.

Callon, M. (1984). Some Elements of A Sociology of Translation: Domestication of the Scallops and the Fishermen of St Brieuc Bay. *The Sociological Review, 32*(1), 196–233.

Chamon, M. (2018). Implied Exclusive Powers in the ECJ's Post-Lisbon Jurisprudence: The Continued Development of the *ERTA* Doctrine. *Common Market Law Review, 55*(4), 1101–1142.

Da Conceição-Heldt, E., & Meunier, S. (2014a). Speaking with a Single Voice: Internal Cohesiveness and External Effectiveness of the EU in Global Governance. *Journal of European Public Policy, 21*(7), 961–979.

Da Conceição-Heldt, E., & Meunier, S. (Guest Eds.). (2014b). Speaking with a Single Voice: The EU as an Effective Actor in Global Governance. *Journal of European Public Policy, 21*(7), 961–1083.

Damro, C. (2006). EU-UN Environmental Relations: Shared Competence and Effective Multilateralism. In K. V. Laatikainen & K. E. Smith (Eds.), *The European Union at the United Nations: Intersecting Multilateralisms* (pp. 175–192). Palgrave Macmillan.

Damro, C. (2012). Market Power Europe. *Journal of European Public Policy, 19*(5), 682–699.

Damro, C., Gstöhl, S., & Schunz, S. (2018). *The European Union's Evolving External Engagement: Towards New Sectoral Diplomacies?* Routledge.

De Zutter, E. (2010). Normative Power Spotting: An Ontological and Methodological Appraisal. *Journal of European Public Policy, 17*(8), 1106–1127.

Degrand Guillaud, A. (2009a). Actors and Mechanisms of EU Coordination at the UN. *European Foreign Affairs Review, 14*(3), 405–430.

Degrand Guillaud, A. (2009b). Characteristics of and Recommendations for EU Coordination at the UN. *European Foreign Affairs Review, 14*(4), 607–622.

Delreux, T. (2006). The European Union in International Environmental Negotiations: A Legal Perspective on the Internal Decision-Making Process. *International Environmental Agreements: Politics, Law and Economics, 6*(3), 231–248.

Delreux, T. (2009). Cooperation and Control in the European Union: The Case of the European Union as International Environmental Negotiator. *Cooperation and Conflict: Journal of the Nordic International Studies Association, 44*(2), 189–208.

Delreux, T. (2014). EU Actorness, Cohesiveness and Effectiveness in Environmental Affairs. *Journal of European Public Policy, 21*(7), 1017–1032.

Dikaios, G., & Tsagkroni, V. (2021). Failing to Build a Network as Policy Entrepreneurs: Greek Politicians Negotiating with the EU during the First Quarter of SYRIZA in Government. *Contemporary Politics, 27*(5), 611–630.

Doussis, E. (2017). *Κλιματική Αλλαγή* [Climate Change]. Papadopoulos Publications.

Doussis, E. (2019). Η Εξωτερική Δράση της ΕΕ για το Περιβάλλον και το Παράδειγμα της Κλιματικής Αλλαγής [The External Action of the EU on the Environment and the Example of Climate Change]. In S. Blavoukos,

D. Bourantonis, & P. Tsakonas (Eds.), *Εξωτερικές Σχέσεις της ΕΕ* [The External Relational of the EU] (pp. 303–320). I. Sideris Publications.

Dreger, J. (2014). *The European Commission's Energy and Climate Policy: A Climate for Expertise?* Palgrave Macmillan.

EUR-Lex. (2020). *Glossary of Summaries: Open Method of Coordination*. https://eur-lex.europa.eu/summary/glossary/open_method_coordination.html. Last accessed: 4 November 2020.

Garben, S. (2019). Competence Creep Revisited. *Journal of Common Market Studies, 57*(2), 205–222.

Garcia, M., & Masselot, A. (2015). The Value of Gender Equality in EU-Asian Trade Policy: An Assessment of the EU's Ability to Implement Its Own Legal Obligations. In A. Björkdahl, N. Chaban, J. Leslie, & A. Masselot (Eds.), *Importing EU Norms: Conceptual Framework and Empirical Findings* (pp. 191–210). Springer.

Gebhard, C. (2011). Coherence. In C. Hill & M. E. Smith (Eds.), *International Relations and the European Union* (2nd ed., pp. 101–127). Oxford University Press.

Gulbrandsen, C. (2013). Navigating from Conflict to Working Arrangement: EU Coordination in the International Maritime Organization. *Journal of European Integration, 35*(7), 749–765.

Haug, C., & Berkhout, F. (2010). Learning the Hard Way? European Climate Policy after Copenhagen. *Environment: Science and Policy for Sustainable Development, 52*(3), 20–27.

Imber, M. F. (1989). *The USA, ILO, UNESCO and IAEA: Politicization and Withdrawal in the Specialized Agencies*. St. Martin's Press.

Ingold, K. (2011). Network Structures within Policy Processes: Coalitions, Power, and Brokerage in Swiss Climate Policy. *Policy Studies Journal, 39*(3), 435–459.

Jørgensen, K. E., Oberthür, S., & Shahin, J. (2011). Introduction: Assessing the EU's Performance in International Institutions—Conceptual Framework and Core Findings. *Journal of European Integration, 33*(6), 599–620.

Juncos, A. E. (2007). *Cometh the 'Hour of Europe', Cometh the Institutions? Coherence and Effectiveness of the EU's Common Foreign and Security Policy in Bosnia and Herzegovina (1991–2006)* [Ph.D. Thesis]. Loughborough: Loughborough University. https://dspace.lboro.ac.uk/2134/7895. 17 March 2019.

Juncos, A. E., & Pomorska, K. (2011). Invisible and unaccountable? National representative and council officials in EU foreign policy. *Journal of European Public Policy, 18*(8), 1096–1114.

Juncos, A. E, & Pomorska, K. (2020). The European External Action Service. *Oxford Research Encyclopedias: Politics*. https://doi.org/10.1093/acrefore/9780190228637.013.1081

Kadelbach, S. (2019). ERTA Case. In *Max Planck Encyclopedias of International Law*. Oxford University Press.

Keukeleire, S. (2003). The European Union as a Diplomatic Actor: Internal, Traditional, and Structural Diplomacy. *Diplomacy and Statecraft, 14*(3), 31–56.

Kissack, R. (2006). *Who Speaks for Europe in the ILO? Member State Coordination and European Union Representation in the International Labour Organization* [Ph.D. Thesis]. London School of Economics and Political Sciences.

Kissack, R. (2007). *European Union Member State Coordination in the United Nations System: Towards a Methodology of Analysis*. European Foreign Policy Unit Working Paper 2007/1.

Klabbers, J. (2009). *An Introduction to International Institutional Law* (2nd ed.). Cambridge University Press.

Knill, C., & Tosun, J. (2009). Hierarchy, Networks, or Markets: How Does the EU Shape Environmental Policy Adoptions within and Beyond Its Borders? *Journal of European Public Policy, 16*(6), 873–894.

Kreppel, A., & Oztas, B. (2017). Leading the Band or Just Playing the Tune? Reassessing the Agenda-Setting Powers of the European Commission. *Comparative Political Studies, 50*(8), 1118–1150.

Kulovesi, K., & Cremona, M. (2014). The Evolution of EU Competence in the Field of External Relations and its Impact on Environmental Governance Policies. In C. Bakker & F. Francioni (Eds.), *The EU, the US and Global Climate Governance* (pp. 81–94). Routledge.

Laatikainen, K. V., & Degrand Guillaud, A. (2010). Two Logics, One Treaty: The Lisbon Treaty and EU Foreign Policy in Brussels and at the UN. *Studia Diplomatica, 63*(1), 3–22.

Larsen, H. (2014). The EU as a Normative Power and the Research on External Perceptions: The Missing Link. *Journal of Common Market Studies, 52*(4), 896–910.

Larsen, H. (2020). Normative Power Europe or Capability-expectations Gap? The Performativity of Concepts in the Study of European Foreign Policy. *Journal of Common Market Studies, 58*(4), 962–977.

Lavenex, S. (2014). The Power of Functionalist Extension: How EU Rules Travel. *Journal of European Public Policy, 21*(6), 886–903.

Lenk, H. (2015). Challenging the Notion of Coherence in EU Foreign Investment Policy. *European Journal of Legal Studies, 8*(2), 6–20.

Lewis, J. (2005). The Janus Face of Brussels: Socialization and Everyday Decision Making in the European Union. *International Organization, 59*(4), 937–971.

Macaj, G., & Nicolaïdis, K. (2014). Beyond 'one voice'? Global Europe's Engagement with Its Own Diversity. *Journal of European Public Policy, 21*(7), 1067–1083.

Manners, I. (2002). Normative Power Europe: A Contradiction in Terms? *Journal of Common Market Studies, 40*(2), 235–258.

Mattocks, K. (2018). Co-ordinating Co-ordination: The European Commission and the Culture Open Method of Co-ordination. *Journal of Common Market Studies, 56*(2), 318–334.

Maurer, H., & Raik, K. (2018). Neither Fish nor Fowl. How EU Delegations Challenge the Institution of Diplomacy: The Cases of Moscow and Washington. *The Hague Journal of Diplomacy, 13*(1), 57–74.

Meunier, S. (2017). Integration by Stealth: How the European Union Gained Competence over Foreign Direct Investment. *Journal of Common Market Studies, 55*(3), 593–610.

Morgera, E. (Ed.). (2012). *The External Environmental Policy of the European Union: EU and International Law Perspectives*. Cambridge University Press.

Musu, C. (2003). European Foreign Policy: A Collective Policy or a Policy of 'Converging Parallels? *European Foreign Affairs Review, 8*(1), 35–49.

Oberthür, S. (2011). The European Union's Performance in the International Climate Change Regime. *Journal of European Integration, 33*(6), 667–682.

Oberthür, S., & Groen, L. (2015). The Effectiveness Dimension of the EU's Performance in International Institutions: Toward A More Comprehensive Assessment Framework. *Journal of Common Market Studies, 53*(6), 1319–1335.

Oberthür, S., & Groen, L. (2018). Explaining Goal Achievement in International Negotiations: The EU and the Paris Agreement on Climate Change. *Journal of European Public Policy, 25*(5), 708–727.

Oberthür, S., & Roche Kelly, C. (2008). EU Leadership in International Climate Policy: Achievements and Challenges. *The International Spectator, 43*(3), 35–50.

Official Journal of the European Commission. (2001). *Commission: European Governance—A White Paper COM(2001) 428 final*, C 278/1.

Panke, D. (2014). The European Union in the United Nations: An Effective External Actor? *Journal of European Public Policy, 21*(7), 1050–1066.

Panke, D. (2017). Speaking with One Voice: Easier Said than Done? The EU in the UNGA. In S. Blavoukos & D. Bourantonis (Eds.), *The EU in UN Politics: Actors, Processes and Performances* (pp. 27–46). Palgrave Macmillan.

Pomorska, K., & Wright, N. (2013). Europeanization and the Common and Security Policy. In C. Bretherton & M. Mannin (Eds.), *The Europeanization of European Politics* (pp. 151–162). Palgrave Macmillan.

Ponzano, P., Hermanin, C., & Corona, D. (2012). The Power of Initiative of the European Commission: A Progressive Erosion? *Notre Europe*, Study & Research 89.

Rama Montaldo, M. (1970). International Legal Personality and Implied Powers of International Organizations. *British Yearbook of International Law, 44*, 111–155.

Rasch, M. (2008). *The European Union and the United Nations: The Functioning and Coherence of EU External Representation in a State-centric Environment*. Martinus Nijhoff Publishers.

Romanyshyn, I. (2015). Explaining EU Effectiveness in Multilateral Institutions: The Case of the Arms Trade Treaty Negotiations. *Journal of Common Market Studies, 53*(4), 875–892.

Russo, E. (2017). Towards an Exclusive Competence of the EU to Conclude Climate Agreements? *European Foreign Affairs Review, 22*(2), 197–212.

Schmitz, L., & Eimer, T. R. (2020). From Coherence to Coheritization: Explaining the Rise of Policy Coherence in EU External Policy. *Globalizations, 17*(4), 629–647.

Schout, A., & Jordan, A. (2005). Coordinated European Governance: Self-organizing or Centrally Steered? *Public Administration, 83*(1), 201–220.

Sjöstedt, G. (1998). The EU Negotiates Climate Change: External Performance and Internal Structural Change. *Cooperation and Conflict, 33*(3), 227–256.

Sjursen, H. (2006). The EU as a 'Normative' Power: How Can This Be? *Journal of European Public Policy, 13*(2), 235–251.

Smith, K. E. (2006). Speaking with One Voice? European Union Co-ordination on Human Rights Issues at the United Nations. *Journal of Common Market Studies, 44*(1), 113–137.

Terpan, F. (2015). Soft Law in the European Union—The Changing Nature of EU Law. *European Law Journal, 21*(1), 68–96.

Thomas, D. C. (2012). Still Punching below Its Weight? Coherence and Effectiveness in European Union Foreign Policy. *Journal of Common Market Studies, 50*(3), 457–474.

Torney, D. (2014). External Perceptions and EU Foreign Policy Effectiveness: The Case of Climate Change. *Journal of Common Market Studies, 52*(6), 1358–1373.

Van Schaik, L. (2010). The Sustainability of the EU's Model for Climate Diplomacy. In S. Oberthür & M. Pallemaerts (Eds.), *New Climate Policies of the European Union: Internal Legislation and Climate Diplomacy* (pp. 251–280). Brussels University Press.

Vanhoonacker, S., & Pomorska, K. (2013). The European External Action Service and Agenda-setting in European Foreign Policy. *Journal of European Public Policy, 20*(9), 1316–1331.

Vedder, H. (2012). The Formalities and Substance of EU External Environmental Competence: Stuck between Climate Change and Competitiveness. In E. Morgera (Ed.), *The External Environmental Policy of the European Union: EU and International Law Perspectives* (pp. 11–32). Cambridge University Press.

Vogler, J. (2013). Energy and Climate Policy. In C. Bretherton & M. Mannin (Eds.), *The Europeanization of European Politics* (pp. 137–149). Palgrave Macmillan.

Wouters, J., de Jong, S., Marx, A., & De Man, P. (2009). *Study for the Assessment of the EU's Role in International Maritime Organizations*. Leuven Centre for Global Governance Studies, Katholieke Universiteit Leuven.

CHAPTER 4

Internal EU Climate Diplomacy: Opening the Black Box

1 Introduction

The systematic posing of climate change targets on international transport is a relatively new direction and goal for the EU. This goal calls for unpacking the institutional foundations, understanding the dynamics between the key actors and for modifying when needed the practices followed within the European system, allowing for the development for an external EU climate diplomacy. Therefore, this chapter analyzes the processes that are taking place within the EU, shedding light on subjects that still remain somehow elusive. Section 2 unfolds the contested issue of EU climate competence regarding international transport policy, while the following Section 3 delves into the core of the internal diplomatic processes that take place within the EU, drawing the picture of the prevailing roles and positions that are dominant. Section 4 illustrates the procedures concerning competence and coordination in establishing a single EU voice. The chapter concludes with Section 5, which portrays the function and the operation of the three *CO*s, namely competence, coordination and coherence, within this process.

© The Author(s), under exclusive license to Springer Nature Switzerland AG 2024
G. Dikaios, *EU Climate Diplomacy Towards the IMO and ICAO*, Palgrave Studies in European Union Politics,
https://doi.org/10.1007/978-3-031-51123-3_4

2 EU Climate Competence: Penetrating in an Additional Policy Field

Environment (including climate change) and transport fall under the shared competence principle, between the EU institutions and the EU Member States. Shared competence, according to Article 2(2)[1] of the Treaty of the Functioning of the European Union (TFEU) and Article 5(3)[2] of the Treaty of the European Union (TEU), dictates that in cases where the Union may exercise its competence in a policy field that European action will deliver better outcomes, the Member States' competence is confined.

Competence in EU climate change policy has been a subject of numerous analyses, as shown in the previous chapter. For as long as it has appeared as a notion that needed policy development in the EU sphere, there has been a struggle between the EU Member States and the European Commission on the type of the competence, with the concept of *subsidiarity* consistently gaining ground (Dahl, 2000). With the passage of time, the institutional setup, through the *initiative right* of the supranational EU institutions, competence has leaned towards and lied with the side of the Commission. The same applies to transport policies, where the EU gradually extended its course of action and gained prominence in a broad array of policies that can regulate any aspect of transportation (Dąbrowski, 2014; Stevens, 2004).

More specifically, the seminal legal acts related to the aviation and maritime emissions are the Directive 2008/101/EC, Regulation (EU) 2015/757 and Directive 2018/410. Starting from the aviation sector and pursuant to Directive 2008/101/EC, the EU included all international and European flights in its ETS system. According to this Directive, the Commission earns the competence to regulate for and control the

[1] When the Treaties confer on the Union a competence shared with the Member States in a specific area, the Union and the Member States may legislate and adopt legally binding acts in that area. The Member States shall exercise their competence to the extent that the Union has not exercised its competence. The Member States shall again exercise their competence to the extent that the Union has decided to cease exercising it competence.

[2] Under the principle of subsidiarity, in areas which do not fall within its exclusive competence, the Union shall act only if and in so far as the objectives of the proposed action cannot be sufficiently achieved by the Member States, either at central level or at regional and local level, but can rather, by reason of the scale or effects of the proposed action, be better achieved at Union level.

GHG emissions from aircrafts that flight within, towards and from the EU (Official Journal of the European Union, 2009). In spite of the 'Stop the Clock' Regulations that were adopted a few years later, Directive 2008/101/EC continued to apply on all the internal European flights, inserting, in this manner, a solid justification for the Commission's action in developing guidelines on how much and in what ways the airlines should mitigate their GHG emissions.

In 2015, the EU adopted the MRV Regulation (2015/757) which handed the Commission the competencies of deciding the ways MRV should take place, as well as the processing of reports submitted by the shipping companies. While this Regulation does not make reference to the reduction of GHG emissions from ships, it paved the way for the Commission to acquire a grasp on the subject and start expanding its authority (Official Journal of the European Union, 2015). A few years later and just before the IMO adopted the Initial Strategy, in 2018, the EU adopted Directive 2018/410 which referred explicitly to the role of the Commission in reviewing 'the progress achieved in the IMO towards an ambitious emission reduction objective, and on accompanying measures to ensure that the sector duly contributes to the efforts needed to achieve the objectives agreed under the Paris Agreement'. It also foresaw that the EU should act towards preparations in order to ensure that GHG emissions mitigation from ships will be implemented. According to several interviewees, both legal documents give the Commission the competence to engage with the GHG emissions from ships, with the Directive enhancing this argument (Interviews no. 5, 6, 13, 35, 40, 52, 54).

In the previous chapters, it was established that the EU is extensively involved in and possesses a significant level of competence on climate change policies, both internally and externally. What remains to be explored is whether this competence has also been extended to the transport sector, in order to provide the EU supranational institutions legal ability to regulate GHG emissions cutting from aviation and maritime activities and to allow those supranational institutions to decide upon the external positions that the EU as a whole should promote internationally. In the case of aviation policy, this is sufficiently embedded in the EU sphere, whereas, in the maritime sector, the institutional setting is still currently under formulation, with significant diverging intakes on the matter. Nevertheless, in order to present a more well-rounded and inclusive image, opinions of the officials of both sectors are going to be presented, not ignoring the fact that deviations are observed in both cases.

In the case of the ICAO, during the interviews of EU Member States and institutions officials participating in the Organization's climate negotiations, three distinct trends surfaced, when attempting to define the level of competence in this case. The first was that the European Commission does have the competence to define the negotiating position and targets which the Member States should then promote internationally (Interviews no. 28, 36, 37, 41, 42), with one of these interviewees—in a rather telling manner—explicitly stating: 'I think that the Member States have transferred absolute authority to the Commission'. The second trend was of a more diplomatic nature. A couple of officials argued that the Commission has the right to initiate proposals that are then discussed with the Member States to agree upon a coordinated position (Interview no. 38, 47), or that there is some competence but not exclusive or comprehensive (Interview no. 44, 45, 51). Finally, the third trend held by a few officials, followed that, in principle, the Commission does not have competence on the subject under concern (Interviews no. 33, 34, 35, 40). The latter argued in favor of the viewpoint that the Member States are the ones who decide and that no position or policy is agreed upon, without their approval.

In the IMO, while the above three trends are evident as well, the balance of the answers is different, proving the statement made by an interviewee that 'we are now in a grey zone' (Interview no. 50). Two of the interviewees claimed that the Commission does not have the competence to regulate GHG emissions from ships (Interviews no. 25, 29), while four others maintained the exact opposite (Interviews no. 26, 46, 49, 52). In fact, one of them argued that although the Union's competence has been challenged by the Member States, this is something that has not been established as an opinion, while another argued that competence exists and is constantly expanding, as the Commission demands to gain more and more ground. The remaining seven interviewees remained undecided and avoided to answer directly (Interviews no. 27, 31, 32, 49, 50) or entirely by falling through the net by claiming that the question at hand is 'complicated and [a] legal question' (Interviews no. 30, 43). Those who accepted to answer underlined the difficult situation that the EU Member States have found themselves in on that matter or referred to the central position the Commission does have in and on the processes.

The striking ambiguity that is observed among the EU Member States and institutions officials regarding the competence and who it lies with, showcases, first, the early phase that the greening of the EU international transport policy (still) is in and, second, the need of systematic exploration of the actions coming from both sides that wish to claim or to preserve power.

Parallel Integration and Functional Spillover

The first mechanism to be examined in order to understand whether (primarily) the European Commission has non-disputable competence in a policy field is that of parallel integration emanating from the EU Member States action. To identify parallel integration in the field of climate change in the international transport sector, two remits should be mentioned. First, action related to climate change in general and, second, action regarding other policy fields related to aviation and maritime sectors. In this point, it has to be recalled that parallel integration takes place outside the EU structure, in intergovernmental negotiations that are aimed in concluding international treaties and where all (or a large number of) the EU Member States participate, affecting thus the internal EU processes.

Parallel integration, to a significant extent, coincides with another mechanism that explains the Union's competence creep, i.e., the functional spillover. Functional spillover takes place through (mostly supranational) action in neighboring policy fields. In climate change and transport policies, which are highly institutionalized, the (independent?) EU Member States action overlaps substantially with the Union's action.

The EU Member States, being committed to multilateral and treaty diplomacy (cf. Delreux, 2011), systematically participate in the development of the international framework regulating GHG emissions. Ultimately, they support the agreements that are formed through these negotiations, i.e., the UNFCCC, the Kyoto Protocol and the Paris Agreement, which have led to the evolution of the European system of mitigating GHG emissions, from the EU. As shown earlier, the European framework builds upon the international one, something that has been proven to allow for the bearing of desired fruits at large.

The participation of the EU Member States at the international climate negotiations had also a significant internal ramification, as it prompted the Commission to come up with the EU ETS, which was established in 2003, on the basis of Directive 2003/87/EC. The Directive gave the Commission the explicit ability to regulate regarding the implementation of its content. It also clearly stated that 'the Commission should, in particular, consider policies and measures at Community level in order that the transport sector makes a substantial contribution to the Community and its Member States meeting their climate obligation under the Kyoto Protocol' (Official Journal of the European Union, 2003), allowing, thus, for the development of a policy and the incorporation of aviation emissions in the EU ETS in the first place.

Therefore, through the action of the EU Member States concerning climate change at the international level, there was and still is a transfer of responsibilities towards the Commission, expanding, in this way, its scope of action in the field of climate change. It is undisputable that governing climate change action, nowadays, encompasses the majority, if not all, of human activities, stretching—by extension—the Commission's function in multiple directions. At the same time, the Commission has also been regulating in all possible sub-policy fields concerning tackling climate change both internally and externally, also pursuant to Article 3(5) of the TEU, which states that the Union 'shall contribute to [...] the sustainable development of the Earth'.

Parallel integration can also be found in the EU Member States action through their membership in the ICAO and the IMO. Apart from two exceptions in the IMO, all the current EU Member States, were Member States of the two IOs before becoming members of the EU.[3] Membership in IOs means constant negotiations, socialization and cooperation with third parties upon specific technical (or political) issues (Magliveras, 2011), that do not always have to do with climate change tackling. For example, in the ICAO, the role of the EU Member States (and the EU as a whole) in promoting security and safety has been incremental and, eventually, extremely influential (Interviews no. 28, 47). It is the European expertise that allows for more entrenchment (as it will be showcased further down in the analysis), and—at the same time—expands the cooperation among the EU Member States through the supranational institutions in the aviation-related field (Barros, 2012).

[3] Germany and Luxemburg became EU members in 1958; the former became an IMO member in 1959, while the latter in 1991. As the United Kingdom was an EU Member State up until January 2020, it is included in timely relevant entries unless otherwise stated.

Accordingly, in the IMO, parallel integration might also be happening through the participation of the EU Member States in the Marine Environment Protection Committee. MEPC regulates usually through legally binding measures a significant array of different polluting gases, such as sulphur or nitrogen dioxide, as well as a wide array of subjects concerning environmental degradation caused by seafaring. For instance, the cut of sulphur emissions from ships was agreed in the IMO in 2008, when obligatory rules for that purpose were adopted. This process brought the EU Member States closer to each other, as they were expected to cooperate in order to implement the global regulation, since the Integrated Maritime Policy of the EU was put in force. The Commission has gained competence regarding this topic (Interview no. 43), through Directive (EU) 2016/802 *relating to a reduction in the sulphur content of certain liquid fuels* (Official Journal of the European Union, 2016b), which falls under the broader issue of air pollution from ships (the same with GHG emissions). To this end, it is worth to mention that the Commission explicitly states in the coordinating papers that distributes prior to the MEPCs that it does have the competence on the sulphur matter (Council of the European Union, 2021a).

Therefore, in this example, both parallel integration and functional spillover happen concurrently. All these actions, from the moment that they are undertaken by the EU Member States, lead to an unavoidable connection to the EU policies. While they might seem disconnected from the climate change negotiations, they built a framework of cooperation which traces back to the internal functioning of the EU. The anterior membership of the EU Member States to the two IOs gave the EU institutions the opportunity to deal with matters beyond their initial scope, by applying the rationale of functionalism, as corroborated by the ERTA Case. The same applies to the international climate change approach which allows for Union-level climate regulations regarding the international aviation and maritime routes, from the moment internal regulations for similar subjects are put in place.

While, in aviation, the EU institutions' action in reducing GHG emissions is crystal clear as all European flights are included in the EU ETS, in the maritime sector the situation is a bit more complex and the development of a climate policy happened more slowly and more gradually. It also started more formally through references in the Communication regarding the Integrated Maritime Policy for the European Union, with a specific mention to the role of the Commission in promoting international efforts to mitigate GHG emissions from ships (Commission of the European Communities, 2007). It was seconded by the White Paper on Transport which put specific mitigation targets on paper (European Commission, 2011). The White Paper on Transport also had reference to aviation attempts in tackling climate change but in much more detail and with a focus on implementing the existing framework (i.e., how to control traffic management to avoid losing competitiveness by CO_2 emissions reductions and how to become a leader in low-carbon fuels for aircrafts) (European Commission, 2011).

Accordingly, the overall EU policy framework on transport and mobility, with the position of the Commission being central in regulating the majority of the policy's aspects, creates spillover effects that transfer competencies to the supranational level, as the more an action is being taken, the more it leads to additional areas that need common regulatory schemes. Last but not least, the promotion of the European Green Deal and the adoption of the European Climate Law has and will continue to have significant effects on the fields of international aviation and shipping. Furthermore, there seems to be wide anticipation that the Commission will gain further competence in the coming years (Interviews no. 25, 27, 31, 33, 34, 38, 39, 42, 46, 52). The European Climate Law expands the role of the Commission in all stages of the policy cycle regarding climate change policy(ies), thus promoting, someone could argue, the European climate action into an *exclusive* competence.

It can be argued, therefore, that the action of the EU Member States within an international framework leads to the extension of the Union competencies in fields such as climate change and transport. Nowadays, as discussion and action on climate change (a central priority of the EU) surpasses and overshadows almost every policy and political aspect, it was a matter of time that this would be extended to (international) transport policies as well. Parallel integration and functional spillover explain the broader framework within which the EU institutions and the EU Member States act. In order, therefore, for the EU to function properly, it develops policies that facilitate the operation of the organizations' function (*effet utile*).

Intergovernmental Bargaining, Institutional Spillover: Soft Law and Soft Political Power

Intergovernmental bargaining as a means for enhancing supranational competence stems from the realization, among the EU Member States, that the results of responsibility transfer to the EU institutions will better serve their interests. This understanding might be also the outcome of the supranational pressures to gain more competence, explaining and claiming that a specific issue would be better managed at the Union level rather than the national one. On the matter of GHG emissions from international transport modes, although a highly institutionalized EU policy with perennial presence exists, we observe that among the EU Member States officials there is not one common understanding of what the Commission's powers are. While for the maritime sector there still is a basis for this uncertainty, this is not to be expected for the aviation sector, at least at the same level. The intergovernmental bargaining takes place at the Council of the European Union (at the Working Party on Shipping and the Working Party on Aviation, respectively) and it is usually led by the proposals of the Commission. It coincides exactly with the coordination mechanism of internal negotiations and will be analyzed in the following section. The remaining paragraphs of the subpart will aim to portray the attempts of the Commission to bend the competence shift in its favor (institutional spillover) by exercising its soft political power.

In climate aviation policy, the institutional constellation on who has the competence to decide on the level of the targets and the negotiating positions to be exported internationally has been defined under Directive 2008/101/EC (which amends the EU ETS Directive, Official Journal of the European Union, 2009) and the continuous practice since then. Therefore, a hard law obligation is in place, contrary to what several EU Member States officials argued. Nevertheless, their claim requires further consideration of the current situation under which the decisions are taken.

On the period prior to the adoption of the CORSIA, the Council Decision (EU) 2016/915 was set into force. The Decision armored the position of the Commission in the development of the global scheme, first, through acknowledging its observer status, second, by outlining its coordination services and, third, through ensuring that the Union position shall be expressed jointly by the Member States (Official Journal of the European Union, 2016a). What this Decision notes additionally, something that maybe is the basis for the various standpoints on the competence issue, is that according to the progress made in the ICAO, 'the relevant Council bodies shall further elaborate their position, in particular as regards the future of the relevant Union legislation within that of ICAO framework'. This passage gives a freedom to the members of the Council of the EU to reshape the position in regard to the international developments, provided that it does not affect the Union's policy. One could argue that the blurry situation emanating from this fluid system is the result of the turmoil emerging from the inclusion of international aviation in the EU ETS and the subsequent 'Stop the Clock' Regulations, which made the EU Member States realize that they need to implement an arrangement that contains checks and balances to control the proposals made by the Commission (Interview no. 35, 36, 40). Thus, the competence of the Commission is described in Article 218(9)[4] of the TFEU on how to approach the negotiations of international agreements, which foresees: 'The Council, on a proposal from the Commission [...] shall adopt a decision [...] establishing the positions to be adopted on the Union's behalf in a body setup by an agreement, when that body is called upon to adopt acts having legal effects'.

Thus, where a hard law obligation to mitigate GHG emissions exists, as is the case in the aviation sector, the process follows the procedures embedded in the EU Treaties, confirming the competence that the Commission has to propose the policy to be followed, but—at the same time—stipulating that it will go through a Council of the EU's decision. To that end, prior to ICAO meetings regarding GHG emissions, information notes are submitted by the Commission to the Working Party on Aviation, which are then discussed and changed where needed. After they are finalized, they are sent to the Permanent Representatives Committee (COREPER) for approval and are distributed to the EU Member States to provide them with information on the position the EU as a whole will support in the ICAO (Council of the European Union, 2019, 2020b).

On the contrary, the situation in the maritime sector is different, as there is not yet a legal obligation to reduce GHG emissions from ships,

[4] The Council, on a proposal from the Commission or the High Representative of the Union for Foreign Affairs and Security Policy, shall adopt a decision suspending application of an agreement and establishing the positions to be adopted on the Union's behalf in a body setup by an agreement, when that body is called upon to adopt acts having legal effects, with the exception of acts supplementing or amending the institutional framework of the agreement.

but only to monitor and report on them.[5] Therefore, Article 218(9) TFEU does not find direct application on the issue at stake, but only an incidental one in the sense that the proposals of the Commission need to be approved by the EU Member States. Nonetheless, according to one of the latest EU nonpapers of the Commission (November 2021) distributed by the Council of the EU to the Working Party on Shipping,[6] in which the coordination procedure takes place, the Commission explicitly states that 'the Union has exclusive competence for GHG emissions in shipping' (Council of the European Union, 2021b: 27). The Commission claims and establishes this authority on the basis of avoiding obstructions to the implementation of the MRV Regulation, the strategy of the EU to include shipping transport in the EU ETS, as well as other policy instruments that emerge through the European Climate Law. 'You could say that we have full competence' a Commission official stated (Interview no. 52).

The efforts of the Commission to establish its competence over the issue of GHG emissions in maritime transport have been present for some years and have been amplifying incrementally, especially since the adoption of the Initial IMO Strategy (Table 1). Based on a series of

Table 1 Legal and other documents invoked by the European Commission to establish its exclusive competence on the climate maritime policy

Renewable Energy Directive (2009/28/EC) and its revision through Directive 2019/2001/EU
EU MRV Regulation (2015/757)
Directive (EU) 2018/410
Climate Diplomacy—Council Conclusions of 18 February 2019
Communication on the European Green Deal of 11 December 2019
Smart and Sustainable Mobility Strategy of 9 December 2020
European Climate Law (Regulation 2021/1119)
Fit for 55 package (14 July 2021)
Research and innovation efforts through the Horizon Europe R&I Framework Program for 2021–2027, through which incentives for achieving zero-emission waterborne transport before 2050 are given

Source Compiled by the author based on information by the Council of the European Union (2021b)

[5] At the moment, several Directives and Regulations are being prepared and expected to be adopted during late 2023 and 2024.

[6] The nonpapers are generally confidential or with limited access, but they are partially accessible to the public on the documents repository of the Council of the EU.

corresponding nonpapers which were circulated prior to MEPC or Intersessional Working Group on Reduction of GHG Emissions from Ships meetings, the Commission reiterated its position that the competence on GHG emissions from the shipping sector falls under its mandate. The EU Member States scrutinize and express general reservations on this matter, requesting the Commission 'to clarify and elaborate in writing for further consideration'. The Commission always responds that these reservations do not serve any specific purpose, as the issue has already been resolved since 2013 through the Working Party on Shipping and calls on the Council to troubleshoot the situation (e.g., Council of the European Union, 2019, 2020a, 2021a, 2021b). The situation remains unresolved to date and might be elucidated by the developments based on the European Climate Law and the Fit for 55 package.

Through the nonpapers and the Council conclusions that (considered as soft law outputs [Di Robilant, 2006]), the Commission instrumentalizes the acquiring of the competence on the matter of GHG emission in the international transport sector, by signposting the competence it has already earned in the internal transport policy.

Although, in the case of the aviation sector, we observe a somewhat better-defined situation, the perceptions among the EU Member States officials remain of outmost relevance for any outcome, as it is not yet well-embedded or even clear on which side the scale is tipping in regard to the competence of who decides about the targets and the negotiating positions.

Further to the use of both hard and soft official documents (which exert soft political power facilitating the institutional spillover), another mechanism that possibly exerts additional soft political power and might prove a defining factor on where the competence of an issue will end up, is that of expertise.

Expertise

It is not only climate change policies that need to align scientific evidence with policy instruments to end up with a feasible plan; transport policies are also considered to have a significant technical nature (Stevens, 2021). Therefore, the expertise of the stakeholders involved in the policy development process is more than necessary to bridge the gap between ideas and action, without though neglecting that knowledge is many times contested within policymaking procedures (Pregernig and Böcher, 2012).

To have better chances to influence these procedures and their outcomes, expertise is more than necessary. Within the context of the European political sphere, expertise tilts the plate of power towards the one who owns it. Thus, in policies where competence is shared, who has the expertise is of outmost importance to also gain the right of decision and lead the creep towards their policymaking arena (Interview no. 32).

Notwithstanding the views that the Commission has the competence to opt for climate targets and positions to be negotiated at the ICAO and the IMO, the prevailing perception among the EU Member States officials, is that their counterparts in the Commission possess strong expertise on GHG emissions regarding aviation and shipping that they themselves lack. Small countries do not have the adequate resources to have experts on such technical and advanced issues and therefore they feel comfortable for the Commission to take the lead (Interviews no. 33, 43). This observation comes to an agreement with the viewpoint from the Commission which claims that the majority of the Member States rely on the Commission to have a position, as only few of the former have substantial, but focused, expertise that has the ability to influence (Interviews no. 40, 47). Further, the fact that the changes being made on the Commission's proposals by the Council of the EU are minor, as a Commission's official argued, showcases that the internal knowledge the Commission has, does influence the EU position (Interview no. 42).

The strong influence that the Commission has on the positions and the consequent action of Member States can be attributed to several correlative functions. First comes the skillful personnel that works on the issues under concern (Interviews no. 26, 28, 34, 36, 39, 43), who gather all the necessary information from different sources (Member States, industries, non-governmental organizations [NGOs], IOs, etc.).[7] Thus, the Commission acquires 'content competence' (Interview no. 44, 51), which eventually leads to the processing of this knowledge and to the decision on which approach would be more beneficial in general and what the negotiating positions should be (Interviews no. 30, 32, 37). Then, the outcome of this process is communicated through coordination, which is seen as the result of the extensive expertise concentration (Interviews no. 30, 32, 34, 36, 38, 39). Coordination takes place among the EU Member States and the Commission, led by the latter.

[7] There are expert groups from the EU institutions which work directly with or within the IOs (Interviews no. 28, 42).

Some skeptical and hardliner voices against the expertise owned by the Commission were also present in the interviews. They claim that the expertise of the EU comes only from the Member States. Their argument is that the Commission officials support the Member States and carry through with the above tasks, but only based on the Member States know-how, as they lack sufficient technical knowledge (Interviews no. 41, 45, 49). While this point of view does not seem to be confirmed in practice, it partly finds support in the literature. Although the Commission is often seen as a technocratic institution, its personnel has grown to be more and more political, pinpointing a trend towards gradual decline of hiring experts (Christensen, 2015a, 2015b).

A final critique on the Commission expertise comes from a more nuanced approach, concentrating on who the experts are and what they advocate for (Interviews no. 26, 44). For a policy proposal to be more influential, it has to be backed up with research, robust evidence and data. The problem, though, is that many of those proposals are conducted by academics who 'lack practicality'. While this makes their acceptance easier because of the authority they bear, especially in contrast to the proposals coming from single Member States, sometimes the goals proposed are unachievable while the practical challenges are kept out, throwing the ball to the industry to find sustainable solutions. The 'request' in order to fix this oxymoron comes in the form of changing up the order of the processes, i.e., the experts to have consultations first with the industry on what is feasible and then build the respective proposals (Interview no. 26).

It is interesting to note that this final argument is not supported by any of the other interviewees who have chosen not to include the industry as a decision stakeholder. However, Commissioner Violeta Bulc, in her interview, gave a picture of the process that put significant weight on the relations between her team and the industry. Bulc narrated her extensive efforts in bringing the industry on board, as well as labor unions and NGOs. In order to provide extra support for this statement, she made notice of extensive meetings she held with shipowners, shipbuilders, aircraft-builders and air carriers, as well as the fact that she travelled to Korea to discuss the promotion of sustainable transportation and get support, e.g., from shipbuilders (Interview no. 48; Delegation of the European Union to Burundi, 2019). Through these relations, her cabinet and team and, by extension, the Commission, constantly gained further and deeper expert knowledge. 'What really made a big difference was the passion of the team and how through meeting after meeting it got more confident, something which made them to be ready in the end, when the day of coordination came. This was fantastic; and this is strength' (Interview no. 48).

Does the European Commission Have Exclusive Competence?

The phantom of climate change creeps over the question of EU competence in the field of international transport. The data converge to the direction that the Commission has indeed the competence to decide which targets and which bargaining positions the EU Member States should negotiate in the IMO and the ICAO. It is the culmination of the processes, based also on the EU Treaties, that explains the competence creep towards the supranational level which allows to argue in favor of the dominant position the Commission gains in the climate aspect of (international) transport policy. In accordance with the principle of subsidiarity and its exploitation by the Commission in policy fields for which it is scientifically proven that transboundary cooperation and solutions are required, the demand of the recognition of the Union's competence is constant and is robustly put into the agenda.

The understanding that the competence regarding climate targets and negotiating goals is settled with the Commission is in agreement with what other scholarly studies have found based on official documents and practice, namely that the Commission has an *exclusive* competence to negotiate climate agreements (Russo, 2017; Van der Mei, 2016). However, this is not applied into practice, where a rather blurry system is in operation, as in the case studies of the IMO and the ICAO. This vagueness is further enhanced by the fact that the Commission is not a full member to the IOs, but only holds an observer status. Thus, the foundational constellation of competence suffers from the institutional reality which ostensibly allows the Commission to acquire an exclusive competence in the field of combating climate change in transport policy but, at the same time, it restrains its ability to exercise it externally.

In more detail, it seems that all the mechanisms analyzed confirm that the Commission has a very important role and that it is gradually acquiring more competence in the field of climate change for international transport. The mechanisms chosen to be studied contribute to two conclusions. Firstly, when it comes to the EU's diplomatic action, all mechanisms should be in harmony, because there would be no way for the European institutions to have a strong position within a system of Member States' remit of diplomatic competence. Secondly, all the mechanisms should work concurrently, otherwise it would be very easy to question the role of the supranational institutions.

Be that as it may, competence as status is in perpetual retrogressive movement between the Member States and the Commission. Especially in an issue such as green transportation, where no clear boundaries exist between their action, either because of the institutional structures or because of the recently very active role that the supranational institutions have, a justification of the position of each actor is highly needed. Therefore, although the Commission seems to be in the privileged position to decide on policy and diplomacy objectives, it is expected to hold extensive coordination meetings with the Member States to achieve an agreement, for two reasons. The first is that a group of the representatives of the Member States involved in the intra-EU and the extra-EU negotiation deny the Commission the competence to set their negotiating strategies, and thus they argue that any position must first be discussed internally to be exported externally. The second reason emerges from the observer status of the Commission in the two IOs, which does not give it the right to vote and it only allows the Commission to take the floor under very certain circumstances. As such, it is up to the EU Member States to promote the positions which serve the European interests, and which have been agreed upon between the involved stakeholders internally, prior to the negotiations in the respective IOs.

Hence, the first work assumption (WA1) that argued that *the EU institutions have the competence to legislate for climate issues concerning the field covered by the IMO and the ICAO and, therefore, determine the negotiating position of the EU Member States and negotiate with both organizations for that purpose*, is partially confirmed.

As it can be derived from the above, the issue of competence remains to a large extent unresolved, with the supranational institutions having a lot of *influence* but much less competence, which allows for the EU Member States to de facto keep a certain, smaller, level of influence on what the targets of the negotiations will be and what bargaining tactics will be pursued. Thus, the coordination process is necessary in order to determine the Union's position in the negotiations taking place at the IMO and the ICAO.

3 Coordination

The ambiguity on the matter of competence when it comes to GHG emissions by international transport opens the door to certain acts of diplomacy that take place within the EU in order for a position to be crafted. This process is called, in the EU jargon, 'coordination' and formally takes place in the Council of the EU, while broader (or informal) procedures also take place in 'the corridors'. Coordination for both the ICAO and the IMO is led by the Commission, which strives to broker an agreement on a common negotiating position and an ambitious—climate—target. Therefore, and according to the usual procedure, the Commission has the initiative to set the content of what the EU Member States shall negotiate at the international level, which then goes through the scrutiny of the latter, in order for the common position to be finalized, adopted and implemented.

In other words, the overarching rationale of coordination lies a lot with what has been discussed so far in relation to the competence issue, as the basis of coordination itself lies with the existing European legislation, which gives the Commission the right to initiate action in the fields of international aviation and shipping. This is accompanied by the soft action within the everyday interplay between the European institutions and the Member States which allows the former to manage the issues at hand and the direction that is going to be followed.

Therefore, several aspects that explain whether the Commission has the competence to deal with climate issues in international transport, i.e., the soft political power, the institutional spillover and the intergovernmental bargain, are also parts of the coordination process that is initiated by the Commission. The Commission, thus, in order to establish its jurisdiction within a specific network that consists of the stakeholders that participate in the internal policy processes of decision-making, sets the agenda in a way that serves its interests and goals.

It has to be noted from the beginning that the intention of the Commission to be deeply involved in the outcome of the ICAO and the IMO climate negotiations is very well established. Apart from the early bold steps in aviation and the more germinal steps in maritime affairs, during the Jean Claude Junker's Commission (2014–2019), a systematic approach was expedited, under the political aegis of the Commissioner of Transport whose mandate clearly stated that 'a modern transport policy can make key contributions to the sustainability of our environment, to help Europe achieve its climate and energy objectives' (Juncker, 2014), with a special mention in the reduction of GHG emissions by the transport sector. The task was intertwined with the competitiveness of the EU and the viability of its growth model. Commissioner Bulc supported that effective and ambitious policies towards the IMO and the ICAO were explicitly within her command, with a goal to bring some real commitments of the global community for truly global transport modes (Interview no. 48). According to her, the achievement of the adoption of the Initial IMO Strategy and the CORSIA was a European success, and it was heavily based on the internal EU coordination with all the relevant stakeholders (Interview no. 48). Both these international outcomes, though, are only the touchstone of climate action by the two respective IOs, and there is no guarantee for their full implementation in the years to come, as several issues remain still open. Continuous diplomatic efforts by the EU seem to be the only way forward for an efficient implementation of the agreements. However, for this to happen, calls for further action internally need to be made, if the EU is to have common positions be crafted and exported internationally.

It is widely accepted that the Juncker Commission was favoring climate-friendly policies in general (Rietig, 2019). Building on that foundation, the current Commission, 2019–2024, under the presidentship of von der Leyen, has been built around promoting climate neutrality as its core political position. The first political guideline to be achieved is the European Green Deal, its primary goal being to make Europe the first climate neutral continent in the world by 2050. In this framework, the maritime and the aviation sector have a prominent position; for both, bold actions are announced to be taken. Using 'big words', von der Leyen in her agenda for Europe, stated:

> I will propose to extend the Emissions Trading Scheme to cover the maritime sector and reduce the free allowances allocated to airlines over time. I will also propose to extend this further to cover traffic and construction. The different systems will have to converge by 2030 if we are to be climate neutral by 2050. (von der Leyen, 2019a)

At the start of her term, von der Leyen formed Commissioners' groups, under which her Vice Presidents supervise based on portfolios relevant to their agenda/mandate, in an overt move towards a more result-oriented scheme. Commissioner of Transport, Adina Vălean, has been placed under the responsibility of Frans Timmermans, 1st Vice President of the Commission who is responsible for the European Green Deal, marking this way a symbolic move highlighting the importance the new Commission is placing on the greening of mobility and transport (European Commission, 2019).[8]

In addition, marking a shift from previous Commissions, Vălean's 'mission letter' has a specific reference to the leading role she will need to perform in the IMO and the ICAO during the negotiations on GHG emissions mitigation (von der Leyen, 2019b). Vălean seems to take these tasks very seriously, stating publicly that we need to 'slap them' with some measures (POLITICO Europe, 2020). While this bold and rather unusual figure of speech was referring to the action of the EU on the international level, it gave a sneak peek on the delineation of the framework regarding the internal EU agenda and how the latter would be set, what goals it would have and the level of ambition it would entail.

Coordination at the Working Party on Shipping (and All Around)

Although the European Commission does not (yet) have the explicit, and on paper, competence to deal with climate change matters in shipping policies, it proceeds to include ambitious targets in the negotiating position that EU Member States should export at the international level. The internal agreement for the supporting of this ambitiousness is a result of the long-lasting negotiations that take place between the Commission, the Member States and sometimes other actors as well, such the European Parliament.[9] As a result, the Commission, ought to find ways which will ensure that all Member States will be in agreement (Interview no. 52). Of course, if EU legislation is in place, the Member States are *sine qua non* obliged to follow it, but for a number of topics discussed in the MEPC of the IMO and regard climate change, the Commission does not have

[8] A similar allocation in groups was also present in Juncker's Commission but with different structure and much stronger inter-Commissioner relations. For example, Commissioner of Transport was mainly under the Vice Presidents of (a) Energy Union, (b) Jobs, Growth, Investment and Competitiveness.

[9] Regarding the process that needs to be followed, technically, for a position to be agreed on, all Member States need to be on board (Interviews no. 27, 31, 43, 46, 52). It is worth to mention, nevertheless, that the Member States cannot decide on the subjects anymore, but only give their consent to the Commission's proposals (Interview no. 32).

a specifically delegated competence (Interviews no. 46, 54; Council of the European Union, 2021b). This, by design, lives room for interpretation, and for reasons of efficiency the Commission proposes coordination (Interviews no. 39, 50).

After the adoption of the Initial IMO Strategy which paved the way for mandatory short-, medium-, and long-term measures to be put in place, the Commission has significantly increased its efforts for this to be done as soon as possible and for the ambitious path that was originally envisaged in the Strategy to be followed (Interview no. 25, 32, 46; Council of the European Union, 2021a, 2021b). To achieve this goal, the Commission pursues rigorous action to ensure that the EU Member States will, first, internally adopt ambitious positions and, second, externally negotiate them at the IMO level. At the same time, the Commission urges the inclusion of the maritime emissions in the EU ETS and the adoption of any other regional framework that would regulate this issue, in case that the IMO decides not to move forward with concluding realistic targets and timelines. This has proven to be a contested subject as the EU Member States' positions diverge (Interviews no. 25, 30), with the interviewees being almost equally distributed to three groups.

The first group is in line with the action propelled by the EU. Its members agree with the level of the Commission's ambition and the negotiating tactics that it follows at the international level. In this category, the Member States go along with the overall EU policy, but also ask for flexibility or exceptions based on their special characteristics (i.e., icebreaker ships, insularity, or landlocked states) (Interviews no. 30, 32, 39, 50). The second group consists of Member States that strive for more ambitious goals, both internally and externally. Those are States that individually have (or discuss to) announce(d) the zeroing of the GHG emissions from their shipping activity before 2050 (Interviews no. 27, 29, 43). Some of this group's members believe that the EU is not ambitious enough.

In direct conflict with this second group comes the third group that is of the opinion that the EU is overly ambitious and promotes unrealistic measures towards the IMO. The members of this group mostly challenge the intention of the Commission to promote a regional regulatory framework of GHG ship emissions and support that the best solution of emissions controlling can only be found within the international IMO structure (Interviews no. 25, 26, 31, 46, 49). Here, the approaches vary from those who reject any ambition or any measure[10] to those who understand the pros and cons of regional measures (or the EU ETS) and believe that this is a first step and a way to push the IMO into a more proactive and efficient direction. The argumentation of this group revolves around the inefficiency of regional solutions, that would first lead to transfer of the shipping activity outside the EU, resulting to the reduction of the competitiveness of the European shipowners and second to carbon leakage. Carbon leakage is one of the main arguments of the internal opposers of the EU action, claiming that it would cause greater damage to the environment. Their position is that the IMO needs to contribute its fair share in mitigating GHG emissions from ships and that global challenges, such as climate change, need global (international) measures. They believe that the Union is very good in wishful thinking, which might be positive in terms of aspirations, but misses the practical aspects of implementation, for which it should respect and listen to its experienced Member States. A recent defeat for these arguments is Brexit, as the British came to realize the needs of European shipping in an abrupt way.

Drawing some early conclusions on the subject, based on the three aforementioned groups, it seems that the positions and role of the Commission are further strengthened (Interviews no. 46, 52). Within a situation blurred enough to be reminiscent of *haar*—the distinctive North Sea fog—and which worsens by the unclear competence condition, the internal negotiations of the EU Member States with the EU institutions prove again to be more than important for the EU to appear coherent at the international level.

The process begins with a proposal by the Commission to the Working Party on Shipping, which entails the targets and the negotiating position that the EU Member States are expected to hold at the IMO. The proposed aspects have to be aligned with the broader policies that have been agreed upon through the ordinary legislative procedure, but in specificities and the(ir) implementation, the Commission has the first saying. Having the initiative to draft a certain policy choice, albeit the need for such a choice to gain a seal of approval by the Member States, the

[10] As indicated by several of the interviewees (Interviews no. 1, 2, 5, 7, 13, 19, 20, 22, 35, 39, 46, 52), the countries with most disagreements are Cyprus, Greece and Malta, but more countries pose different kinds of reservations during the process of internal negotiations.

Commission exercises its power to influence (Interviews no. 26, 27, 39, 49). This influence takes several formats. The Commission always invokes previous political acts and EU legislation to persuade the Member States to follow its lead (Interviews no. 30, 32) and it works together with the ambitious and the willing of the Member States to promote its agenda (Interviews no. 25, 29, 39).

If the aforementioned process fails, the last steppingstone the Commission utilizes is the 'threat' to level up the negotiation venue, i.e., to take the discussion to the COREPER where the national ambassadors to the EU participate (Interviews no. 43, 46, 52). At the COREPER, the dynamics are different because the bargain is political and a wide range of topics is discussed, not only related to maritime affairs, and Member States do not have the same influence as they do in the Working Party on Shipping, since any issue might become part of a package deal in a wider political negotiation. Therefore, to avoid this process and the consequent tensions it may cause, the tendency is to try to agree on a position at the Working Party as much as possible. This procedure, of course, puts pressure on the minority; it does not matter if a Member State is more or less ambitious if it ends up isolated. In cases like the one just discussed, it is to be expected that if a Member State insists and persists in an argument, after a certain amount of time, no one will listen to its positions and it will be treated as a 'pain in the neck', with whatever that entails. Ultimately, this state will have to give up, as the clock will remain ticking and a possible non-unanimous outcome might undermine this state's negotiating position within the EU. Although some find the internal negotiations intense (Interview no. 26, 14, 49), others feel that the atmosphere within the Working Party is good (Interview no. 32, 50) and, as such, the officials often feel obliged to follow the position to keep the good negotiating climate (Interview no. 13, 43).

While the general tendency is supporting the ambitiousness, there are several issues that trouble the Member States. First of all, some believe that it should be the Member States that decide upon the negotiating positions and the targets and, therefore, have more flexibility to act in a manner that will benefit the ultimate goal, which is the reduction of ship GHG emissions (Interviews no. 25, 26, 27, 29, 43, 46).

The reasons for this tendency can be both internal and external. Internally, several Member States would wish to advocate in favor of more ambitious targets for both the EU level and the international one (Interviews no. 27, 43). Here, the Commission plays a slightly negative role as it waters down their ambition and their efforts in a bid to reach common ground with the rest of the Member States who do not wish to follow. Externally, there is a fear that the high ambitions promoted robustly by the EU will limit the flexibility of the negotiations, because overly ambitious targets and aggressive international approaches might put the desired outcome at risk (Interviews no. 25, 46).

There is a general agreement on how active, and for some even pressing, is the European Commission on persuading the EU Member States on the matter of climate change towards the IMO. The interaction between the Commission and the Member States takes multiple forms, i.e., informal or formal bilateral and multilateral meetings. The same applies within the Working Party on Shipping, as well. Over the last few years, the pattern of these meetings has changed and a particular increase in their number has been observed. The Commission has begun to prepare the work and the common positions drafts by conducting more and more informal meetings with the EU Member States and launching more and more expert groups. Through this process, it attempts to deeper understand the situation and to grasp the intentions of the key Member States that are willing to be more active than others (Interviews no. 25, 27, 50, 52).[11]

After the adoption of the Initial Strategy, when the short-, medium-, and long-term measures needed (and some still need) to be agreed upon, the IMO negotiations move forward through a system of submissions, where individual IMO Member States submit proposals or already developed policies to be discussed by the MEPC. Submissions in the IMO serve as proposals or comments on the agenda items of the respective meetings; they are submitted prior to the meetings and the negotiations take place based on them. Utilizing the standing within the IMO of the Member States that wish to play a role in the negotiations, the Commission (which cannot engage with all the aspects of reducing GHG emissions from ships, due to its Observer-only status) proceeds

[11] This fact marks a deviation from what was going on until before the adoption of the Initial IMO Strategy, when the regular process of coordination was followed (Interviews no. 8, 9, 10, 16, 20, 23). Thus, an intensification of the process is observed since mid-2010s.

to cooperate with the latter in an array of fields.[12] On that fact, it has to be mentioned that according to a number of the interviewees, the Commission would rather be able to submit the respective proposals and to stipulate the content itself, but the EU Member States react and insist that the submissions should be made jointly for all the EU Member States (Interviews no. 13, 25, 27, 46). This slight turbulence in the relations of the Commission and the Member States can be seen in the cases when the former finds itself in a difficult position because the Member States have co-sponsored submissions with third parties (Interview no. 52), something that allows them to retain their privilege and showcase their superiority within the system of the IMO. In order for this irregularity to be evened out, the issue of representation has recently been brought up before the CJEU.[13] The question of the specific case (Case C-161/20) is whether the Commission has the exclusive competence in areas governed by the IMO as well, and therefore when decisions are about to be taken, pursuant to Article 3(2)[14] of the TFEU which gives the Commission the exclusive competence for international agreements that affect its internal responsibilities, the Commission is the body to decide what the position of the EU as a whole will be.

Through all these processes, it is no surprise that a prevailing view is that the Commission has a central and controlling role regarding climate change tackling from international shipping. Nonetheless, there are some voices that argue that the Member States' positions are not overdriven, that the coordination drafts are heavily modified and that in the case of reservations more discussions take place in order for an agreement to be achieved (Interviews no. 27, 29, 30, 31, 39). Conversely, there are several diverging viewpoints that claim that their objections and challenges on the Commission's proposals are 'effectively' silenced by the latter and no room for lobbying of the national positions is left. It is worth to mention that the changes that the Member States want to make in the coordinating

[12] For example, Germany prepares the joint submissions for all the EU Member States regarding fuel standards, Denmark does the same for market-based measures, Finland for working arrangements, etc. (Interview no. 52).

[13] See Berg (2021) for an initial legal comment on the Case C-161/20.

[14] The Union shall also have exclusive competence for the conclusion of an international agreement when its conclusion is provided for in a legislative act of the Union or is necessary to enable the Union to exercise its internal competence, or in so far as its conclusion may affect common rules or alter their scope.

drafts that include the final positions are being recorded as reservations in footnotes and they are not extensively discussed during the coordination process (Interviews no. 14, 22, 25, 26). In addition, the tactic of 'bombarding' the Member States with coordinating papers and the overrun of the process with 'a thousand meetings' is seen, by some, as way to 'loose a sense of direction' of what actually is going on. This 'process' was also emboldened during the COVID-19 pandemic, where the meetings were taking place online. Many times, there was no time to prepare or to speak or even to participate on some occasions. This has resulted in the Commission presenting decisions that have never been heard by the Member States before. Summarizing the situation of how the Member States feel, abruptly put, 'the Commission sometimes steps back and takes into account some of the Member States' views; *it is not purely a dictatorship*' (Interview no. 26).

Coordination at the Working Party on Aviation (and Beyond)

The institutional structure of the coordination in aviation issues at the European level is a lot different from what is taking place in shipping affairs, and a lot of coordination before meetings in the ICAO takes place, including outside the European institutions in the strict sense. Apart from the EU body in which the internal negotiations take place, i.e., the Working Party on Aviation at the Council of the EU, there are also additional institutions that include extra-EU countries at the European continent. The most prominent of these bodies is the European Civil Aviation Conference (ECAC), which today consists of 44 countries located in the European continent, 27 of which are the Member States of the EU.[15] ECAC's mission coincides with the goals of the ICAO in terms of harmonization of common practices but is focused on Europe in particular. It also aims to promote understanding with other regions around the world.

The EU's action towards the ICAO mostly takes place through the work that happens at ECAC, enhancing, this way, the power and the standing of the European region at the international level (Interview no. 36). For the EU, the participation in ECAC bears some advantages, as it offers a venue for testing the strategies which it aspires to promote at the

[15] ECAC was founded in 1955 in Strasburg by 22 countries in the European continent.

ICAO in its surrounding countries, as well as a means to gather expanded regional experience (Interview no. 28).[16] For climate change and environmental issues, in particular, there is also a specific group that functions within the ECAC framework, the European Aviation and Environment Working Group (EAEG), which is co-chaired by an ECAC Member State and a representative of the Directorate General for Mobility and Transport (known as DG MOVE) of the European Commission.[17] The EAEG consists of the European CAEP members, observers and ad hoc rapporteurs of the rest of the ECAC members to the CAEP or experts, as well as representatives by the European Union Aviation Safety Agency (EASA) and EUROCONTROL.[18] When the discussion concerns the reduction of the GHG emissions by international aviation, the EAEG meets in an extended setup, which caters to the (pan-)European coordination efforts. For some of the EU Member States officials who participate in all the respective *fora*, the heavy work in coordinating a European position which will be then negotiated at ICAO takes place at the EAEG (Interviews no. 36, 38, 40).

Therefore, the coordination process that takes place at the European level is much more complex and contains different levels of cooperation before its outcome travels on the other shore of the Atlantic. But what happens within the EU? The internal EU negotiations are of outmost importance for a common position to be exported externally, whether this is at the ECAC/EAEG or at the ICAO. After all, in the last few years, the ECAC usually follows the EU position; a position that has been first worked on within the Union (Interview no. 44).

The landscape of how the Commission performs the coordination efforts with the EU Member States is again following the normal procedure, but it seems to be much less politicized and uncontested in contrast to what happens in the Working Party on Shipping. Here, the adherence to more ambitious climate targets and stringent ICAO strategies is a

[16] It also creates problems to the position that the EU wishes to promote because countries with very different needs in aviation participate in the ECAC, such as Ukraine (Interview no. 41).

[17] The Commission's representative does not co-chair the sessions when the issues on the table fall out of the EU competencies. For more information on EAEG, see European Civil Aviation Conference (2022).

[18] 'EUROCONTROL is a pan-European, civil-military organisation dedicated to supporting European aviation' (EUROCONTROL, 2022).

widespread belief and goal, and the support to the EU position is unconditional (Interviews no. 33, 34, 36, 37, 38, 42, 44, 45, 51). Diverging intakes, though, appear through the internal negotiations mostly when States want to protect or to promote their aviation industry, targeting the way the prospect policies will be implemented and financed rather than fighting the ultimate goal of having the GHG emissions significantly reduced by international aviation or having the timeframe of policy implementation altered (Interviews no. 33, 34, 36, 38, 44). Maybe this is because of the longer and much more advanced action of the Union when it comes to aviation affairs, which has established a solid and widely accepted understanding of the climate action the EU institutions want to pursue. Another, more prosaic, approach that was observed in the interviews supports that this is the result of the existing EU law; the EU Member States have previously decided that an ambitious climate policy towards international aviation is to be adopted and, based on the principle of sincere cooperation and the unity of international representation, they have to support a common EU position externally (Interviews no. 42, 44, 47).

Nevertheless, the Member States attempt to contain the thrusting entrepreneurship of the Commission (Interviews no. 42, 47) through 'hard fights' on the proposals that the latter initiates (Interview no. 34). The drafts distributed by the Commission to the Member States are thoroughly reviewed and only if all the Member States come to an agreement, then a position is concluded.[19] This happens, naturally, through coordination and mostly for cases when the competence of the Commission is not very clear (Interview no. 42). In ICAO issues, more specifically, the definition of the positions is taken in the Council of the EU, according to Article 218(9) TFEU. Nevertheless, 'EU Member States cannot act alone in GHG emission policies internationally and they have to follow an EU position' (Interview no. 47). For climate aviation issues, as put by a national representative: 'internal coordination is continuous – coordination, coordination, coordination; this is the mantra of the European Commission' (Interview no. 45). The Commission is deeply involved in and leads the coordination process, which is constantly active and online (Interview no. 34, 37, 47).

[19] For example, the Commission made a proposal on airline environmental labelling (euobserver, 2021) which was not accepted by the Member States and the Commission is still occupied on it (Interview no. 34).

For the Member States, the Commission is too far ahead in terms of climate ambition (Interview no. 36, 41, 45). Commission officials are not modest in their goals, but at the same time they are not eager to pass their position; the final decision comes through negotiations. As all Member State officials 'get the chance to take the floor' (Interview no. 37), the Commission contends for the maximum possible outcome in the internal negotiations, in order for the external EU position to be as much as ambitious it can get (Interview no. 34). However, although the Commission in private conversations boldly maintains what it wants to be the goal, on an official level, it utilizes diplomatic language and is polite in asking for the dialogue to go to a specific direction (Interview no. 37).

In some cases, the Member States have to discuss one of the many agenda items of the Working Party on Aviation, because one or more of them does not agree to what the Commission proposes, or this proposition is unrealistic (Interview no. 44). In these cases, the most effective persuasion tool of the Commission is legislation and funding. For most of the reluctant Member States, the divergence in ambition is usually caused by questions on finances and economics. To that end, discussions on investments and jobs take place during the coordination process, to make sure that no restrictions will be posed on businesses or no huge impacts will national economies suffer. If that 'fear' is taken off the table, usually through contributory benefits that are not publicly known (Interview no. 34), it is much easier to conclude to a common—ambitious—position (Interview no. 33). Notwithstanding, it is also because the Member States understand the advantages of a common position and do not want to push back and impede the entire agreement (Interview no. 44).

In addition, according to Commission officials, not a single Member State aims to obstruct the process in any way; it is not about persuading the Member States to support the EU position externally, because they are part of the EU, and this relation is regulated by the EU law (Interview no. 42, 47). It might happen that some Member States are more active[20] than others and might take the lead to some issues and be more influential in shaping the EU's position (Interview no. 40), but the Commission's proposals have been usually accepted with only minor changes (Interview no. 42). The above seems to confirm the viewpoint that the Commission is more powerful of what it appears, using all the knowledge and expertise

[20] For example, Germany, Netherlands, France (Interviews no. 36, 40, 41).

that it has on top of taking into account the lobby work that is constantly in operation (Interview no. 37). As it is to be expected, this also raises concerns regarding the far-reaching competencies that the Commission might be acquiring through this process; for at least one interviewee, the tactic of conjuring up agreements made years ago and using them to pursue current matter must be treated with caution (Interview no. 44).

Internal Coordination: Restrictive and Beneficial

Internal coordination can be both restrictive and beneficial to the EU's climate diplomacy in the field of international transport. It is also both restrictive and beneficial for the Commission. The common feeling, though, for the Member States leads to the conclusion that internal coordination is more restrictive for their position, rather than more beneficial.

It is confirmed through the above that the role of the Commission is strong within the internal negotiating procedures. It pushes significantly for specific positions and goals, that frequently do not conform with what the EU Member States want, and it takes advantage of all the means it is allowed to, within the boundaries of the institutional framework, and it extends its action where the margins are unclear. More specifically, as the Commission does not have the right to be directly and fully represented at the two IOs, it employs its coordination prerogative to bound the Member States to promote the position it would have otherwise negotiated itself. Acknowledging the fact that the Member States have diverging interests in the issues at stake, the Commission predominantly frames the subject as urgently important and it approaches the Member States with the highest ambitions possible, in order to end up to a common denominator that would satisfy all the involved parties, or at least the majority of them.

A lurid example of the Commission's framing is the division between the international and the European scheme. In more detail, the Commission threatens the EU Member States with unilateral action (Interviews no. 14, 26, 52), which will culminate into the establishment of a binding European system, a possible development that none of the Member States seem to find attractive. The Commission states that it wishes international schemes to be in place and be sufficiently ambitious in order to contribute to sustaining the Earth's temperature to not rising over 2 °C, but in the unlikely failure of concluding such schemes, the Commission declares that it is ready to adopt its own rules to apply within the EU. Although the EU Member States seem to maintain a position that restrains the action of the Commission on how the negotiations will take place, there is a fear that if the negotiations will not bear the anticipated fruits, the Commission will initiate strict and highly ambitious measures to be implemented in the EU that the Member States will not manage to refute. Such development would be based, of course, on the will of some Member States to support such mechanisms, especially the ones who feel that they have to lower their standards in order for a common EU position to be adopted. According to an interviewee, 'the Commission must first listen to what the Council wants and therefore have its proposals be within the extent possible and be realistic. The Commission does not go against the flow and the co-configuration between the two bodies is incredible. Nothing appears out of the blue' (Interview no. 35). Thus, maybe the reason that the Commission appears so combative, apart from its own competencies and extensive work on the issue of combating climate change, lies with the support it enjoys by a significant number of the EU Member States. This would not be easy for the latter to admit publicly, especially from the moment that they usually prefer showing off any achievements as theirs, rather than as a collective success (Interview no. 35).

Nevertheless, in the current scheme of things, in order for the Commission to promote its desired positions externally, it needs to first get the EU Member States to agree, as they are the agents officially transferring the EU positions to the two IOs. Thus, the Commission extensively outreaches to persuade the Member States to follow the EU line and ensure that no deviations will occur during the international negotiations. It could be argued that the Commission has the competence to coordinate, the *coordination competence*, which also emerges through its deep knowledge of what the problems are and how things work. For a Member State official (Interview no. 30), the expertise of the Commission equals to its ability to coordinate the EU's action and to setting the agenda.

The above procedures continue to strengthen the arguments of the Commission for the acquisition of full competence and for expanding its jurisdiction in issues regarding climate change in international transport policies. Notwithstanding, in spite of the long and continuous practices that give prominence to the role of the Commission, the EU Member States still have a certain level of influence over the outcomes of the internal negotiations keeping some checks and balances to the incremental power of the former.

The relations between the supranational and the national level regarding the climate policy of aviation seem more institutionalized and smoother in contrast to what is observed in the maritime sector. While in aviation the majority of the officials find the fact that the Commission executes a coordinating role which leads to an enhanced EU position during the international negotiations quite helpful, in shipping this is not exactly the case. The reasons behind this divergence could be—at least to some extent—the following: first, the inclusion of the aviation emissions in the EU ETS in 2008, as well as the harsh diplomatic experiences gained after the failure in imposing the scheme on international flights, might have defined the roles between the Commission and the EU Member States in a compromise that allows both sustaining good relations with the ICAO and third parties in general. Moreover, the utilization of Article 218(9) TFEU bounds, at the same time, the Commission to rely on the agreement of the Member States and the Member States to follow invariably the position agreed through internal coordination procedures. Second, the fact that there is yet no similar legal obligation for the shipping emissions, but the Commission strives robustly to develop one, creates significant tensions between the supranational and the national levels. While everyone ostensibly agrees that ships should reduce their GHG emissions, there are lot of disagreements on the ways this will be achieved or on the followed processes. There is also a growing concern on the role the European Climate Law will have on the freedom of the Member States and the stated ambition of the Commission to make the obligations more stringent for both sectors. Nevertheless, the overall feeling is that the aviation sector is more prepared to accept and negotiate the consequences of the European Climate Law, in contrast to the shipping sector that seems paralyzed in front of the rapid political changes that took place during the last few years and the gains of ground of the Commission in its field.

Thus, the two work assumptions concerning internal coordination are confirmed. *Internal coordination is necessary to strengthen EU climate competence and the European Commission performs it to strengthen its position in the EU climate governance structure* (WA3). Additionally, *internal coordination is needed in order to reach an agreement on the negotiating position and strategy* (WA2), especially from the moment that WA1 was not fully confirmed, and the Commission is not able to execute exclusive competence on climate issues in transport policy. Consequently, the Commission needs to perform extensive negotiations with the EU Member States to, first, enhance its role in the internal level and, second, to appear coherent at the external.

4 Internal Coherence: Establishing a Single EU Voice

What does coherence mean for the European context? Does it mean that the EU Member States and the EU institutions speak with a single voice in their external relations or that they just support a common position, albeit with diverging (?) voices? These questions alone build up a puzzle that tends to confirm that coherence does not actually resemble an explanatory concept but rather a wishful notion of how the EU should appear. In addition, in the cases under examination, the understanding of the key stakeholders and participants to the negotiations of what 'a single voice' means and whether it is beneficial for the EU aspirations can give us several explanations on how this 'notion' is perceived practically. Once more, it has to be mentioned that coherence is the result of the coordination process, whereas it remains intertwined with the nature of the competence in the issue at stake. Thus, the views of the EU Member States' officials frequently view the coherence notion through the lenses of either the result of the coordination process or the level of the EU competence (Fig. 1).

Taking into account all answers by the EU Member States' officials that participated to the IMO and the ICAO climate negotiations, three groups of quite clear-cut views emerge in the answers. There were EU Member States' officials that immediately responded 'no' to the question on whether they agree to the EU having a single voice, others who said that a single voice is definitely advantageous, and some who diplomatically refrained from giving a sharp answer by arguing in favor of coherent position but not of a 'single voice', or by saying that sometimes it is good to have a single voice but sometimes it is not. These three groups, for indicative reasons, are labeled accordingly 'no', 'yes', and 'so and so'. It is interesting to note that we find officials that work in both the maritime and the aviation sectors in all three categories of responses. This allows for a dual explanatory function, i.e., to examine their reactions all together and make a broader observation of how Member States officials perceive the Union as a (single) actor and its external action in general.

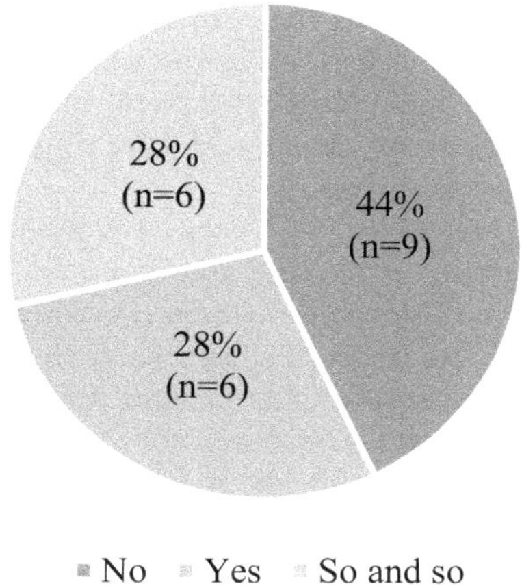

Fig. 1 Reply to interview question "Do you agree with the idea that the EU should speak with a 'single voice' in its external relations, even if some of its Member States interests diverge?" ($N = 21$)

Although the spontaneous answers for 'no' were slightly more in number, no strong percentage is recorded, nor is this common answer solid in essence. Only five out of 20 argued in favor of a definite denial of the single voice of the EU. The first interviewee argued that only the Council Presidency should speak on behalf of the EU, not the EU Observer, and that when there is not a coordinated position the Member States should go separately and alone (Interview no. 30). The second accused the Commission that it pursues a specific agenda that it is not always suitable to the Member States' interests and thus it should maintain its position only as an Observer with the Member States leading the negotiations. For this official, the previous experience in these Organizations has shown that when the single voice is absent, the EU is more successful (Interview no. 37). The third and the fourth 'nos' said the same thing with different words. They claimed that the single voice impedes the 'coalitions recipe', which will lead the EU eventually to be incapable of negotiating anything or with anyone.[21] They also blamed the Commission that it wanted to have a common position for every item during the negotiations, ending up distributing a hundred pages documents before them. According to this interviewee, the Commission should confine

[21] This was also supported by other interviewees in the sense that the single voice can be beneficial, but it depends on how strict the negotiating position is and how much it takes interests of the EU Member States into account (Interview no. 25, 45).

itself only to the subjects it has competence and prioritize the issues of importance, such as the overall goals and not the technicalities of their implementation (Interview no. 46). Putting it in a cruel manner, the other Interviewee said '[the single voice] is a mistake; *we are a democracy*' (Interview no. 45).

The fifth 'no' departs from different origins. The argument in this opinion is twofold and functions as well as a bridge between the negative and the positive stances on coherence: first, EU diplomacy works despite the common position, so there is no need for the process of a unified voice; second, the *de jure* design of the IOs which counts one vote for each state would require major changes for the EU to become a single actor, without losing its power (Interview no. 39). This argumentation also coincides with what the Commission's officials support, namely that it is the numbers that matter and that the purely intergovernmental nature of the IOs work in favour of the EU interests (Interviews no. 42, 47).

Based on this approach, the majority of the interviewees argue in favor of a coordinated position that allows for the EU as a whole to appear as a coherent actor. Within this cast, and according to the diverging initial answer each gave (i.e., 'yes', 'no', 'so and so'), the officials justified their answers via different approaches. The ones who responded 'no' claimed that *single voice* and *coordinated position* serve two different purposes with the latter being more effective and offering more advantages (Interviews no. 26, 36, 38, 43). The answers taken from the 'so and so' group, pinpointing the need of a coordinated position, were stemming from the same direction, although they were expressed in different words (Interviews no. 27, 29, 50, 51). This could be attributed to the fact that by using varied language and different angles or by putting emphasis on an array of aspects allows for better maneuvering during international negotiations. This is also very important in the early stages of a negotiation where the EU Member States might experience trust issues if they speak with a single voice. It is a firm belief of the participants that better solutions can be achieved if they have room to be flexible, depending highly on the issue at stake, the stage of the negotiations and whether the latter are formal or informal.

Conversely to the two previous approaches, the rest of the interviewees favored the single voice, although not without expressing skepticism on the competence issue and the prevalent role of the Commission in the coordination process (Interviews no. 25, 28, 31, 34, 41, 44). The reasons behind defending the single voice notion sketch a holistic image of how the EU as a whole performs at the international level when it comes to a global issue.

First of all, because climate change is a transboundary challenge, there is a need of international action and of a robust single position in order to make your voice heard. The EU Member States could not in any case regulate the reduction of GHG emissions alone and, in this way, have a positive influence in tackling the effects of climate change. To this end, the fact that the EU consisted of 28 Member States has a special weight during the negotiations. It was common for the interviewees to wonder what Germany or France would be able to achieve if they were alone, without the backing up of the rest of the EU Member States. Therefore, the single voice seems a suitable way forward in being effective and successful when the opposition comes from 147 in IMO and 165 in ICAO. 'Recent history has provided several examples on how strong the influence of Europe may be when its positions are expressed as single' (Interview no. 28), especially when serving the environmental protection objective (Interview no. 25). Adding to that, the credibility that the EU enjoys when it appears coherent leads the Union to look for a ***modus operandi*** that aims for the creation of a single position to be exported internationally.

Coherence (and single voice) can be simply identified as 'being existent' or not. Thus, the findings above confirm that coherence resembles to ***more of an idea***. This conclusion emerges because both mechanisms that measure coherence, namely ***negotiating determinacy*** and ***coheritization*** happen mostly through the internal coordination process and the transferring of further competencies to the supranational level. Negotiating determinacy refers to the structuring of the negotiating position that confines the EU Member States to follow a specific line. Coheritization occurs when the Commission actively advocates for the positions it wishes to establish. Work assumption (WA4), which expects that ***coherence is based both on the status of the EU competence on the subject and on the internal coordination results*** is, therefore, confirmed. Although the competence of the supranational institutions in the international transport policy regarding climate change is still in question, the Commission—through implementing coherence—finds the opportunity to request the establishment of a 'single EU voice'. Further, it is common that when the EU has appeared as one actor in a particular issue, it is using this instance as a leverage to convince the Member States that it is for their own benefit and reputation to continue acting coherently (Interviews no. 25, 44). As mentioned earlier, several Member States' officials believe that through its 'single voice' the EU becomes a credible global actor.

It is apparent that 'coherence' and 'single voice' are perceived as both an internal and an external aspect of the EU's international action by those who perform it, as well as by the scholarly literature. For some few, the EU as a whole should not appear as coherent as it is possible, as the interests of the individual EU Member States should be heard at the international level. For the majority, though, notwithstanding the different nuances in the answers they gave, the understanding of the power of the EU when it speaks with a single voice or, at least, with a coordinated position, is clear and always works for the benefit of the EU.

5 Concluding and Discussing the Three *CO*s: Dynamics in Flux

The three *CO*s (competence, coordination and coherence) craft the essential stipulations for the EU as a whole to appear and perform as a normative power at the international level. This chapter shed light on their meaning and on how they intertwine when the EU comes to interact with other IOs, namely the IMO and the ICAO, in the field of climate diplomacy.

*Co*mpetence is a concept that stems from the legal basis of the EU Treaties, which determine the roles that each actor within the European sphere acquires. In the case of climate and transport policies, like the ones at hand, when it comes to competence, things are not so simple. The two policy areas are governed through the shared competences between the EU Member States and at the supranational level on the basis of the principle of subsidiarity. In both fields, the Commission exercises its right to regulate and pull the competencies towards its jurisdiction, thus being able to perform exclusive functions. This fact, though, in combination with the complex institutional framework that exists for the representation of the EU as a whole within the IMO and the ICAO, creates tensions between the Commission and the EU Member States. Systematically, the Commission claims that it has the exclusive competence to decide on climate matters that pertain to transport policies and, thus, it has the right to define the positions and the targets that will be negotiated at the international level. Contrarywise, the EU Member States, as the only ones able to officially represent an EU position to the two IOs, often argue that it is their prerogative to decide what they are going to support through their national delegations.

On a practical level, at the ICAO, the EU's action is covered by Article 218(9) TFEU, as any decision at the ICAO regarding the reduction of climate change will have consequences in the internal legal order of the EU (following the Directive 2008/101/EC which includes aviation emissions in the EU ETS). Nevertheless, this is not the case in the EU–IMO relations, as no EU law is in place controlling GHG emissions mitigation that come from ships yet, thus the aforementioned Article does not find application. The reason behind the absence of such a legal obligation concerning ship emissions could be due to past shortcomings of the EU. Back in 2012, the Union had attempted to impose its ETS on international aviation, an attempt that ultimately failed. This led to a tactical retreat of the EU on the issue in regard to aviation and to approaching the second pillar of international transport with more caution, in order not to repeat the same strategic mistakes. Thus, the smaller steps with which the EU proceeded in regulating GHG emissions from ships have not yet culminated in any demarcated arrangement on how the EU should work, creating a contested situation in which, lately, the Commission strives for gaining as much power as possible on the maritime climate field.

For both cases—at different levels of intensity—the Commission methodically attempts to acquire further powers, following the procedures that were analyzed in this chapter, to expand its climate competence fan. This general strategy creates friction and diverts public and institutional attention from the cases at hand. A telling passage from the interview with a seasoned European Commission officer with very long experience in transport policies refers to 'legalistic measures [which, being put forward by the Member States to overcome the supranational prevalence] take attention from the subject' (Interview no. 52). These legalistic measures are usually put in the discussion through the internal *Co*ordination process, which is the ordinary procedure in the functioning of the EU, especially when shared competence is employed and no agreement has been (or can easily be) reached. In addition, it seems that such frictions have a long-standing history inside the internal politics of the EU, and they might also have a positive function. Morgenstern-Pomorski states very accurately that 'the attempt by the Member States to keep the Commission at arm's length, yet still relying on its input and recognizing the need for coordination with it, formed part of the contradictions of

European political cooperation from the very beginning' (Morgenstern-Pomorski, 2018: 25). The aforementioned internal coordination is highly relevant in the cases of the IMO and the ICAO, as the Commission cannot represent the EU in their respective General Assemblies.

Legal(istic) measures are, however, also put forward by the Commission to secure its principal role in all the procedures. From what it has been discussed so far and taking stock of the vague situation that covers the competence issue on the matters under examination as explained, what emerges through the analysis is that the Commission possesses a (the) *coordination competence*. Coordination competence can be defined as the possession of knowledge on how to coordinate a group on a specific issue, for which the coordinating actor has gathered all the relevant and important information and can decide or propose which is the path to go. This is confirmed through a number of interviews during which, Member State officials complained that the Commission tries to have a saying on everything, as it would be better to restrain itself and focus only on subjects where it has a clear competence (Interview no. 45, 46). Again, this leads to a contested landscape, which results in the even more disputed function of speaking with a single voice on behalf of the EU. While in the ICAO the issue of *co*herence is settled through Article 218(9) TFEU, this is still not the case for the IMO.

In the case of the IMO, in order to clear the institutional horizon and to put an end in the diverging inputs, the Commission appealed to the CJEU to ask for a ruling on the matter of 'on behalf' of whom the EU Member States can speak in the IMO. The preliminary opinion of Advocate General Szpunar in late November 2021 (Case C-161/20) drowns the hopes of the Commission to win any gains against the Member States in the most emphatic way. The ruling of the CJEU (which was published in April 2022) followed a largely similar rationale (Hadjiyianni & Herodotou, 2023).

Through the analysis, we witness that a potential strong opposition of the Member States in any of the discussed mechanisms will most likely result in the prevalence of more intergovernmental aspects, at the expense of a strong and influential role of the EU in other international fora. This fact leads to the conclusion that the totality of the mechanisms that are being applied with the goal to promote EU climate diplomacy need to be happening concurrently and cumulatively, in order for the EU's power to reach the level of influence it has the ability to attain in the supranational level; an ability that will be undermined in case any of these procedures is left out, leading the EU climate diplomacy to suffer.

However, to that end, the evidence suggests that a convergence is taking place among the different mechanisms. It seems that, conversely, the Commission 'goes through the cracks', as put by Violeta Bulc, that the Member States leave *unguarded*, either because they do not want to take action on a specific issue or because they prefer the action to be taken by and at the supranational level.

It is *sine qua non* that the (initial) powers delegated to the Commission enjoy the agreement of all the Member States. This acceptance (and commonplace) means that the system that is developed is stable and solid and cannot/does not operate according to the wishes of any single Member State or group of Member States that, in a certain point of time, do not like what they had agreed upon earlier. Eloquently put by an interviewee, the idea of the *à la carte* EU is obsolete, and it is argued that the Member States ought to remember what they agreed on and to 'not complain' every now and then for issues that have already been settled (Interview no. 44).

The all-pervading ambivalence that has emerged from the interviewees regarding competence issues and internal coordination (that often takes place with the goal to impose supranational positions concerning an external negotiation within traditionally intergovernmental institutional frameworks [in our case the IMO and the ICAO]), enhances and emboldens the argument for the concurrent examination of the mechanisms proposed, leading to a more clear identification of competence and explaining coordination, with an emphasis on the latter.

Theorizing, by analyzing coordination, the perceptions and the point of views of the immediately involved and concerned stakeholders are unpacked and the process allows for a better image of the procedures taking place behind closed doors to be formed. Moreover, the value that each or all the Member States put on an issue, while showcasing the dynamics that are shaped between different institutional actors, is also depicted. The overlap among the mechanisms becomes crystal clear, confirming that in order to gain deeper and better understanding of the functioning of the European *diplomatic* apparatus in its broader conceptualization, the need for including the revealing personal views and the interpersonal relationships of the actual officials that are involved in the discussions is of utmost importance.

By extension, the same need to understand the operation of the diplomatic processes themselves will continue to find application, regardless of any changes that will come in the near or more distant future. A good example for this would be the case of the implementation of the newly established European Climate Law. The role of the officials, both from the Commission and the Member States, will gain further prominence in the near future as the dynamics of the relations between the supranational and the national levels are expected to somehow shift along the long road to climate neutrality goal by 2050. According to the 'Fit for 55' package that the Commission communicated to explain how the interim target of 55 per cent reduction of GHG emissions will be achieved by 2030 (European Commission, 2021), a series of legal and substantive actions are being unfolded, something that will robustly impact the European course of action in maritime and aviation climate policies, among all other aspects of climate policy. The following Table 2 gathers the proposals of the 'Fit for 55', for which legislative action is currently under way (e.g., the development of sustainable fuels in the aviation and maritime sectors is being discussed in the European Parliament [2022]).

Table 2 'Fit for 55' proposals concerning climate maritime and aviation sectors

<table>
<tr><th>Maritime</th><th>Aviation</th></tr>
<tr><td>Extend the EU ETS to maritime emissions</td><td>Advance the EU ETS in regards to aviation emissions by abolishing free allowances and incorporating the CORSIA requirements into the European system</td></tr>
<tr><td>FuelEU Maritime: production of cleaner and sustainable maritime fuels</td><td>ReFuelEU Aviation: developing more sustainable fuels for aviation</td></tr>
<tr><td colspan="2">Alternative Fuel Infrastructure Regulation: assuring the availability of alternative fuels for all modes of transportation</td></tr>
<tr><td colspan="2">Revision of the Energy Taxation Directive to include, among others, maritime and aviation sectors in order 'to align the minimum tax rates for transport fuels with the EU climate and environmental objectives, while mitigating the social impact'</td></tr>
<tr><td></td><td>Implementation of the Single European Sky, which is expected to reduce GHG emissions of airplanes by 10%</td></tr>
<tr><td colspan="2">Ensure continuous cooperation with the EU's international partners on the respective policy issues (including the IMO and the ICAO)</td></tr>
</table>

Source European Commission (2021)

Interestingly, all these proposals can be perceived as the capstone action of the Commission to regulate the reduction of GHG emissions from maritime and aviation sectors, while—concurrently—ensuring a principal role for itself in governing all policies emerging hereinafter. This might also be seen as an active attempt by the Commission to reserve the more powerful position within the new legal system under development for itself, by affirming the apprehension of some of national officials that in the next years will find the Commission with more (or even exclusive) competence in regulating GHG emissions from ships and aircrafts.

But a possible conundrum is emerging. What will the ramifications of the approval of these legislative procedures be and how will the Commission bring together its potential power increase with the fact that it cannot be directly represented at the institutional structure of the IMO and the ICAO? In addition, with the ruling of the CJEU and the Spuznar rationale possibly prevailing in the coming years, apart from a whole new chapter opening in the future implementation of the ERTA doctrine and the functionality of the Union's operation, the following paradox will also emerge: the Commission will have stronger jurisdiction in climate transport policy but no opportunity to externalize it as a common European position, without the voices of its Member States.

List of Interviews

Interview no. 1, Advocate of the industry of an EU Member State, phone call, 31 March 2019.
Interview no. 2, Advocate of the industry of an EU Member State, phone call, 12 April 2019.
Interview no. 5, EU Member State official, phone call, 16 April 2019.
Interview no. 6, EU Member State official, phone call, 25 April 2019.
Interview no. 7, EU Member State official, phone call, 25 April 2019.
Interview no. 8, EU Member State official, phone call, 25 April 2019.
Interview no. 9, EU Member State official, phone call, 30 April 2019.
Interview no. 10, EU Member State official, phone call, 30 April 2019.
Interview no. 13, EU Member State official, phone call, 3 May 2019.
Interview no. 14, Advocate of the industry of an EU Member State, phone call, 7 May 2019.
Interview no. 16, EU Member State official, email exchanges, 10 May 2019.
Interview no. 19, European Commission official, phone call, 22 May 2019.
Interview no. 20, EU Member State official, phone call, 22 May 2019.
Interview no. 22, EU Member State official, phone call, 24 May 2019.

Interview no. 23, EU Member State official, phone call, 27 May 2019.
Interview no. 25, EU Member State official, virtual call, 2 November 2021.
Interview no. 26, Advocate of the industry of an EU Member State, virtual call, 10 November 2021.
Interview no. 27, EU Member State official, virtual call, 11 November 2021.
Interview no. 28, EU Member State official, virtual call, 16 November 2021.
Interview no. 29, EU Member State official, virtual call, 17 November 2021.
Interview no. 30, EU Member State official, virtual call, 17 November 2021.
Interview no. 31, EU Member State official, virtual call, 17 November 2021.
Interview no. 32, EU Member State official, virtual call, 18 November 2021.
Interview no. 33, EU Member State official, virtual call, 24 November 2021.
Interview no. 34, EU Member State official, virtual call, 26 November 2021.
Interview no. 35, European Parliament official, virtual call, 26 November 2021.
Interview no. 36, EU Member State official, virtual call, 30 November 2021.
Interview no. 37, EU Member State official, virtual call, 1 December 2021.
Interview no. 38, EU Member State official, virtual call, 1 December 2021.
Interview no. 39, EU Member State official, virtual call, 3 December 2021.
Interview no. 40, European Commission official, virtual call, 3 December 2021.
Interview no. 41, EU Member State official, virtual call, 3 December 2021.
Interview no. 42, EU Member State official, virtual call, 3 December 2021.
Interview no. 43, EU Member State official, virtual call, 6 December 2021.
Interview no. 44, EU Member State official, virtual call, 7 December 2021.
Interview no. 45, EU Member State official, virtual call, 7 December 2021.
Interview no. 46, EU Member State official, virtual call, 8 December 2021.
Interview no. 47, European Commission official, virtual call, 8 December 2021.
Interview no. 48, Violeta Bulc, European Commissioner of Transport (2014–19), virtual call, 20 December 2021.
Interview no. 49, Advocate of the industry of an EU Member State, phone call, 3 January 2022.
Interview no. 50, EU Member State officials, virtual call, 4 January 2022.
Interview no. 51, EU Member State official, virtual call, 10 January 2022.
Interview no. 52, European Commission official, virtual call, 24 January 2022.
Interview no. 54, European Commission official, virtual call, 10 February 2022.

References

Barros, X. (2012). EU Counterterrorism and Aviation Security: Supranational Rules but Intergovernmental Politics? *European Foreign Affairs Review, 17*(Special Issue), 53–69.
Berg, C. (2021). The External Representation of the European Union in the International Maritime Organization: A Question of Labelling rather

than of Competence. *EU Law Analysis: Expert Insight into EU Law Developments*. http://eulawanalysis.blogspot.com/2021/12/. Last accessed: 30 March 2022.

Christensen, J. (2015a). Recruitment and Expertise in the European Commission. *West European Politics, 38*(3), 649–678.

Christensen, J. (2015b). The Organization of Professional Expertise in the European Commission. *Politics and Governance, 3*(1), 13–25.

Commission of the European Communities. (2007). *Communication from the Commission to the European Parliament, the Council, the European Economic and Social Committee and the Committee of the Regions: An Integrated Maritime Policy for the European Union*. COM(2007) 575 final.

Council of the European Union. (2019, September 10). *Union Position for the Reply to ICAO State Letter on CORSIA Sustainable Aviation Fuels (SL 19/55)—Endorsement*, 11886/19.

Council of the European Union. (2020a, June 19). *75th Session of the IMO Marine Environment Protection Committee (London, 30 March–3 April 2020)—EU Coordination*. 6419/2/20 REV 2 LIMITE.

Council of the European Union. (2020b, June 10). *Union Position on the Reply to ICAO State Letter ENV 6/6–20/53: Proposed Amendments to the ICAO Document 'CORSIA Sustainability Criteria for CORSIA Eligible Fuels', reference in Annex 16, Volume IV*, 8728/20.

Council of the European Union. (2021a, May 11). *76th Session of the IMO Marine Environment Protection Committee (10–17 June 2021)—EU Coordination*. 8277/1/21 REV 1 LIMITE.

Council of the European Union. (2021b, November 15). *77th Session of the IMO Marine Environment Protection Committee (22–26 November 2021)—EU Coordination*. 13527/1/21 REV 1 LIMITE.

Dąbrowski, M. (2014). Transport Policy: EU as A Taker, Shaper or Shaker of the Global Civil Aviation Regime? In G. Falkner & P. Müller (Eds.), *EU Policies in a Global Perspective: Shaping or Taking International Regimes?* (pp. 130–148). Routledge.

Dahl, A. (2000). Competence and Subsidiarity: Legal Basis and Political Realities. In J. Gupta & M. Grubb (Eds.), *Climate Change and European Leadership: A Sustainable Role for Europe?* (pp. 203–220). Kluwer Academic Publishers.

Delegation of the European Union to Burundi. (2019). *EU Commissioner for Transport Meets with Seoul Mayor*. https://eeas.europa.eu/delegations/burundi/63685/eu-commissioner-transport-meets-seoul-mayor_zh-tw. Last accessed: 28 March 2022.

Delreux, T. (2011). *The EU as International Environmental Negotiator*. Ashgate.

Di Robilant, A. (2006). Genealogies of Soft Law. *The American Journal of Comparative Law, 54*(3), 499–554.

euobserver. (2021, February 22). EU Preparing Airline Eco-Labels for 2022. *Euobserver*. https://euobserver.com/tickers/151001. Last accessed: 4 February 2022.

EUROCONTROL. (2022). *About Us*. https://www.eurocontrol.int/about-us. Last accessed: 29 May 2022.

European Civil Aviation Conference. (2022). *European Aviation and Environment Working Group (EAEG)*. https://www.ecac-ceac.org/activities/environment/european-aviation-and-environment-working-group-eaeg. Last accessed: 4 February 2022.

European Commission. (2011). *White Paper: Roadmap to a Single European Transport Area—Towards a Competitive and Resource Efficient Transport System*, COM(2011) 144 final.

European Commission. (2019). *Decision of the President of the European Commission of 1 December 2019 on the Creation of Commissioners' Groups*, P(2019) 4.

European Commission. (2021). *Communication from the Commission to the European Parliament, the Council, the European Economic and Social Committee and the Committee of the Regions—'Fit for 55': Delivering the EU's 2030 Climate Target on the Way to Climate Neutrality*, COM(2021) 550 final.

European Parliament. (2022, February 28). *ReFuelEU Aviation and Maritime: ENVI Exchange on Opinions*. https://www.europarl.europa.eu/committees/en/refueleu-aviation-and-maritime-envi-exch/product-details/20220214CAN64971. Last accessed: 1 April 2022.

Hadjiyianni, I., & Herodotou, A. (2023). The EU's Representation in International Organizations: Case C-161/20 *Commission v. Council (International Maritime Organization)*. *European Foreign Affairs Review, 28*(1), 30–52.

Juncker, J. C. (2014, November 1). Mission Letter: Violeta Bulc, Commissioner for Transport. *European Commission*. https://wayback.archive-it.org/12090/20191120070534/https://ec.europa.eu/commission/commissioners/sites/cwt/files/commissioner_mission_letters/bulc_en.pdf. Last accessed: 3 February 2022.

Magliveras, K. D. (2011). Membership in International Organizations. In J. Klabbers & Å. Wallendahl (Eds.), *Research Handbook on the Law of International Organizations* (pp. 84–107). Edward Elgar Publishing.

Morgenstern-Pomorski, J. H. (2018). *The Contested Diplomacy of the European External Action Service: Inception, Establishment and Consolidation*. Routledge.

Official Journal of the European Union. (2003). *Directive 2003/87/EC of the European Parliament and of the Council of 13 October 2003 Establishing a Scheme for Greenhouse Gas Emissions Allowance Trading Within the Community and Amending Council Directive 96/61/EC (Text with EEA relevance)*, L 275/32.

Official Journal of the European Union. (2009). *Directive 2008/101/EC of the European Parliament and the Council of 19 November 2008 amending Directive 2003/87/EC so as to include aviation activities in the scheme for greenhouse gas emission allowance trading within the Community (Text with EEA relevance)*, L 8/3.

Official Journal of the European Union. (2015). *Regulation (EU) 2015/757 of the European Parliament and of the Council of 29 April 2015 on the Monitoring, Reporting and Verification of Carbon Dioxide Emissions from Maritime Transport, and Amending Directive 2009/16/EC (Text with EEA relevance)*, L 123/55.

Official Journal of the European Union. (2016a). *Council Decision (EU) 2016/915 of 30 May 2016 on the position to be taken on behalf of the European Union with Regard to the International Instrument to Be Drawn Up within the ICAO Bodies and Intended to Lead to the Implementation from 2020 of A Single Global-Based Measure for International Aviation Emissions*, L 153/32.

Official Journal of the European Union. (2016b). *Directive (EU) 2016/802 of the European Parliament and of the Council of 11 May 2016 relating to a reduction in the sulphur content of certain liquid fuels (codification)*, L 132/58.

POLITICO Europe. (2020). *Opening interview with EU Commissioner Vălean | POLITICO's Sustainable Future Summit*. https://www.youtube.com/watch?v=Gp1_HW1ebyk. Last accessed: 9 February 2022.

Pregernig, M., & Böcher, M. (2012). Normative and Analytical Perspective on the Role of Science and Expertise in Environmental Governance. In K. Hogl, E. Kvarda, R. Nordbeck, & M. Pregernig (Eds.), *Environmental Governance: The Challenges of Legitimacy and Effectiveness* (pp. 199–219). Edward Elgar Publishing.

Rietig, K. (2019). The Importance of Compatible Beliefs for Effective Climate Policy Integration. *Environmental Politics, 28*(2), 228–247.

Russo, E. (2017). Towards an Exclusive Competence of the EU to Conclude Climate Agreements? *European Foreign Affairs Review, 22*(2), 197–212.

Stevens, H. (2004). *Transport Policy in the European Union*. Palgrave Macmillan.

Stevens, H. (2021, September 6–8). *The Management of Mature Policies—The Case of Transport*. Paper presented at the UACES 51st Annual Conference.

Van der Mei, A. P. (2016). Case Note: EU External Relations and Internal Inter-Institutional Conflicts: The Battlefield of Article 218 TFEU. *Maastricht Journal of European and Comparative Law, 23*(6), 1051–1076.

von der Leyen, U. (2019a). *A Union That Strives for More—My Agenda for Europe: Political Guidelines for the Next European Commission 2019–2024*. https://ec.europa.eu/info/sites/default/files/political-guidelines-next-commission_en_0.pdf. Last accessed: 3 February 2022.

von der Leyen, U. (2019b, December 1). *Mission Letter: Adina Vălean, Commissioner for Transport*. https://ec.europa.eu/commission/commissioners/sites/default/files/commissioner_mission_letters/mission-letter-valean-2019-2024_en.pdf. Last accessed: 3 February 2022.

CHAPTER 5

Performing Normative Climate Diplomacy at the International Level

1 Introduction

It is a firm belief of the interviewees that the European action significantly influences the adoption of ambitious climate policies and targets in the IMO and the ICAO (Fig. 1).

It is noteworthy, though, that in breaking down the answers that each EU Member State official gave according to the international organization they engage with, a divergence between approaches emerges. For the officials taking part in the proceedings of the ICAO, the sentiment that the EU is an important player in promoting climate change rules is stronger than for the ones in the IMO (Figs. 2 and 3). Nevertheless, the overall result remains relatively high in favor of the EU's influence, and it is worth to mention that all the interviewees recognize that at least some impact of the Union on the international transport sector regarding the reduction of GHG emissions is measured.

According to the conceptualization by Manners (2002), a norm in order to be diffused needs to follow a certain procedure. This section will follow this conceptualization, as well as the structure of (Table 3 in Chapter 3), as to give prominence to the features that are in concordance with what Manners proposes and test whether the diplomatic action of the Union is instilled with normative practices.

© The Author(s), under exclusive license to Springer Nature Switzerland AG 2024

G. Dikaios, *EU Climate Diplomacy Towards the IMO and ICAO*, Palgrave Studies in European Union Politics,
https://doi.org/10.1007/978-3-031-51123-3_5

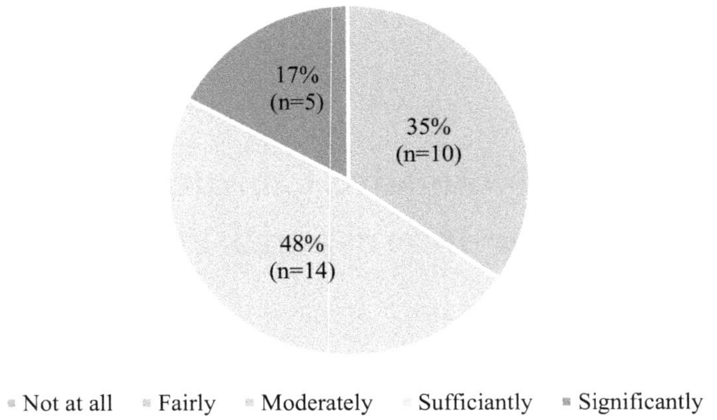

Fig. 1 Combined answers for both IOs to the question "How much do you think that the EU (as a whole) influences the IMO/ICAO to adopt ambitious climate policies and targets?" ($N = 29$)

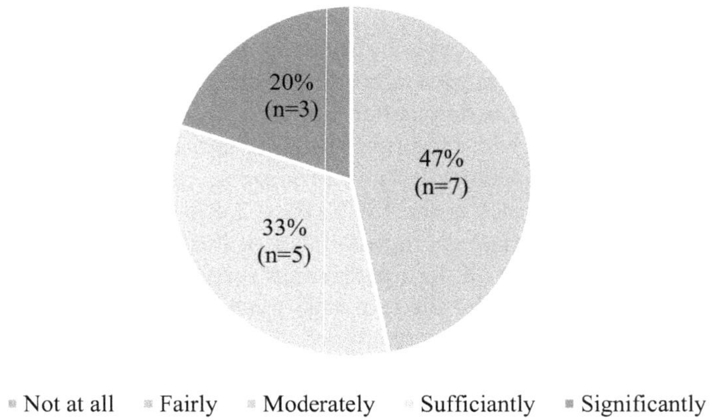

Fig. 2 Reply to interview question (IMO) "How much do you think that the EU influences the IMO to adopt ambitious climate policies and targets?" ($N = 15$)

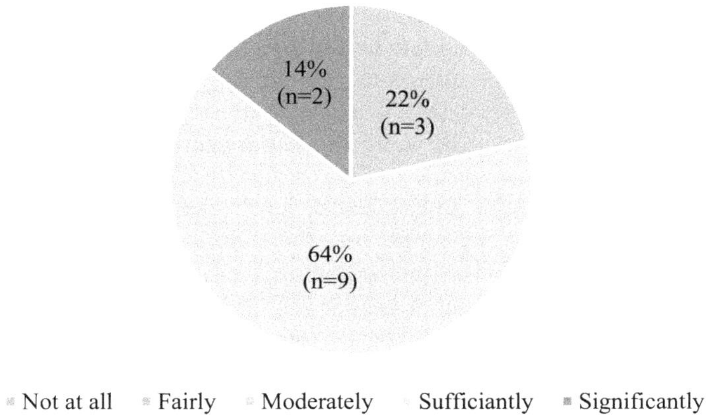

Fig. 3 Reply to interview question (ICAO) "How much do you think that the EU influences the ICAO to adopt ambitious climate policies and targets?" (N = 14)

2 Contagion (or Where Strategy Meets Unintentionality)

By now, it has been made evident that the EU promotes climate measures both at home and abroad, something which creates a certain tendency in the international arena of often treating the EU as a climate leader. This fact could also lead to unintentional influence towards third parties that come across the EU's path. But does this happen? Does the EU unintentionally affect the adoption of international measures regarding the reduction of GHG emissions or is there always a strategic plan to this end?

Almost all of the interviewees agreed that the EU's climate attitude towards the IMO and the ICAO (and, maybe, also in general) is strategically planned and intentional. Roughly half of them, however, seemed to recognize that the existence of one of the most comprehensive and developed systems of tackling climate change also leads to unintentional influences towards third parties. As this statistical finding is rather significant, it is also very telling that the majority of the Member State officials who work on maritime emissions argued that the action of the EU towards the IMO is very much intentional and well organized, while the majority of the officials participating in the aviation negotiations acknowledged the EU's overall influence at the international level as more prevalent, whether it is intended or unintended.[1] Once more, this fact showcases how the governance situation in the EU maritime affairs is more contested, in contrast to EU aviation affairs, as the ICAO participants seem to have better fathomed the Europeanization of the aviation policy.

Focusing back on the international level, the EU unintentionally diffuses its climate policy by supporting the technical issues regarding the implementation of the GHG emissions mitigation strategies. As the EU is able to demonstrate that some measures are feasible through its expertise, it significantly influences the international course of action (Interviews no. 26, 31, 33, 34, 36, 38, 39, 43, 44). Therefore, having the EU being so active at the negotiating table by pushing regulations and showing the way with its internal policies, an aspiration/expectation of something happening without bargain materializes, as, according to the interviews, third parties will understand that doing something for climate is progressive and 'the way forward' (Interviews no. 28, 29, 34, 39, 51). For the Europhiles and the optimists, the EU's decisions are the driving force of what others will do (Interviews no. 35, 42). For the less romantic, the EU is a major block of countries and a major economy, which can only be compared to those of the USA and China; therefore, third parties that wish to work along with the EU and to benefit from such a cooperation, know in advance that any agreement will contain several pro-climate aspects (Interviews no. 26, 28, 30). An additional point in this unintended, but welcome, influence comes from the fact that the EU promotes policies for the *greater good*,[2] vesting its advocacy with a moral cloak and, in a way, virtue-signaling its action. Consequently, it is not easy to reject (at least publicly or loudly) what the EU supports, as this might

[1] To be precise, eight out of 11 interviews that argued in favour of the existence of a strategic planning come from the maritime sector. Conversely, seven out of 11 interviews arguing that the EU action is both intentional and unintentional come from the aviation sector.

[2] The assumption that the EU is *a force for good*, whether it is coming from academics or practicioners, is obsolete since the early 2000s. The literature has been critical to this idea, pinpointing that those actions do not usually follow a moral or ethical rationale (Brown 2002 at Jørgensen & Laatikainen, 2004). The *force for good* concept has also been scrutinized in relation with the self-perception of the EU being a normative power (Sjursen 2006a, 2006b).

have broader negative consequences on other diplomatic relations (e.g., naming and shaming) (Interview no. 49).

For the rest of the EU Member States officials (the ones who do not believe in unintentionality), the action of the EU is always strategically planned, very well organized, with openly known ambitious targets and a clear goal to convince others to implement mitigating GHG emissions strategies on their (economic) activities. This does not mean that these officials neglect that the EU's action does have unwitting effects, but they give greater emphasis on the need to pursue a specific (diplomatic) tactic in order for the European position to prevail during the negotiations (Interviews no. 25, 36, 37, 38, 44).

Further, internal EU action often sounds an alarm to the respective IOs prompting them to proceed in actions, in order to hinder regional measures by the former. Instead, the IMO and the ICAO seem to prefer the devising of global solutions over regional strategies. This is the case because of their fear that the expansion of regional measures might end up in making them irrelevant or, worse, obsolete, a development that both IOs strive to avoid at any cost (Interviews no. 26, 39, 50).

Conversely, the EU is itself applying two different approaches in its relations with the IOs at hand when it comes to climate change. In the case of the ICAO (possibly after its failed attempt to include international aviation in the ETS), it has accepted that a global solution is the more feasible option. On the other hand, as the IMO still struggles to practically adopt any even slightly ambitious climate policy, the Union is flexing its muscles, threatening with the creation of stricter regional measures.

However, it is worth to note that the EU's aforementioned attitudes towards the two IOs are dynamic. As pointed out earlier, Commissioner Vălean has been very vocal about her attitude towards the ICAO, while the emergence of the European Green Deal, the Fit for 55 with its shipping provisions, as well as the MRV Regulation a few years earlier, forewarned the IMO that action will be taken by the EU (Interviews no. 26, 43, 46).

Nevertheless, the interviewed officials argue in favor of a global solution, particularly since the shipping and the aviation sectors are global in nature, and some of them doubt that clear EU impacts can be spotted in the international policy outcomes, as influences come from many different actors and directions (Interviews no. 31, 39, 44). For some, however, including certain European officials (Interviews no. 31, 35, 44, 45, 52), the EU is alone in combating climate change and for that reason it needs to push for climate action and to intensify its role as the climate leader: 'otherwise no other country or block will do anything' (Interview no. 45); 'no one has come to tell us that what we are doing is fine' (Interview no. 52).

Anecdotally, one interviewee shared a similar view through a simile: 'It is like when you are at school, and you are the best student in the class; no one really likes you. You give all your homework to your co-students, but they do not do it the same way as you or at the same level. The EU does push things forward, but the rest of the world is not looking forward to embracing the great EU ideas as a divine inspiration; they just grumpily follow along' (Interview no. 27).

The perception of the unintended effects of the EU's action is part of the wider understanding of the long presence of the EU's goal to persuade the IMO and the ICAO—as well as other IOs—to engage in combating climate change (Kulovesi, 2012; Lin, 2017). While the expectation stemming from *contagion* is that the EU norms would unintentionally diffuse ideas to third actors, the understanding of the participating stakeholders in the negotiations tells us a different story. It is as clear as day that they believe that the EU's diplomatic activity is very well-defined and has very specific goals that it aspires to achieve. Only by coincidence, some unintentional effects take place, mostly in the sense of what others expect that the EU will advocate for. On the contrary, it is the EU that needs to lead by its example (Interviews no. 31, 44) in order for international climate action to take place, especially by the IMO and the ICAO.

Maybe it is *contagion* that provokes the heated discussion on whether the character of the EU's action is normative or strategic, but—in any case—two questions arise: First, are strategic and normative two mutually exclusive concepts? Second, does a norm has to be diffused unintentionally for the action to be characterized normative? Rejecting this division that has occupied a significant part of the literature,[3] practice in the climate change field shows that in order for the EU climate *norm* of mitigating GHG emissions to be successfully embraced by the rest of the world, it needs to be *strategically* exported. The process could also have some unintentional influence, but it mostly emerges as a consequence of a well-defined strategy.

[3] See Chapter 2, Sect. 2, herein.

3 Informational Diffusion

Based on the literature, the most common factor to appear in the EU–IMO and the EU–ICAO relations is informational diffusion, i.e., the stated policy initiatives and declaratory communications. This does not come as a surprise, since the easiest way to assess the influence that one actor has on another is through the study of official documents. The EU started regulating emissions from the shipping and aviation activity, within its geographical jurisdiction, long ago (Kopela, 2017; Veno, 2007). The reason for the development of these systems was the 'dissatisfaction' with the IMO and the ICAO (Lindenthal, 2014; Van Leeuwen, 2015; Van Leeuwen & Kern, 2013) for their reluctant and ambiguous action, which is often judged as slow and with dubious results (Martinez Romera & van Asselt, 2015; Nengye & Maes, 2012; Oberthür, 2006; Shi, 2018; Interviews no. 14, 15, 16, 35, 40, 52). This is the case as, on the one hand, it is recalled that, while the IMO recognized climate change as a problem in the late 1990s, it only adopted some timid measures against it in 2011 (Harrison, 2017; Shi, 2018; Shi & Gullet, 2018). On the other hand, in the ICAO the situation was clearer, due to the inclusion of international aviation in the EU ETS from early enough and prior to the EU's systematic diplomacy push towards the Organization. Failing to establish (or impose) an ambitious policy, the EU retreated from its initial position and began drafting and publishing documents in order to build a grid of proposals to highlight that an implementable system was possible for the international level as well (Gehring & Robb, 2018; Lindenthal, 2014).

The following Table 1 gathers an indicative list of pivotal official documents that different European institutions have produced, showcasing their intention to engage systematically with climate diplomacy (with a focus on the two IOs), as well as their will to establish stringent policies in contrast to the ones that are currently in force, with the introduction of additional ramifications on the international level as well. The time span of the Table 1 ranges from the adoption of the CORSIA for the ICAO and the Initial Strategy for the IMO to 2021,[4] covering the most active period of the EU attempts to develop a concrete climate diplomacy towards the IMO and the ICAO.

[4] For an extended and comprehensive list of such documents, see European Commission (2022a, 2022b).

Table 1 EU official documents concerning climate diplomacy with IMO and ICAO

Document title and type	*Year of publication*	*Published by*	*Addressed to*
Council Conclusions on Climate Diplomacy	2016	Council of the EU	IMO & ICAO
A European Strategy for Low-Emission Mobility	2016	European Commission	IMO & ICAO
Council Conclusions on Climate Diplomacy	2018	Council of the EU	IMO & ICAO
Directive (EU) 2018/410 amending Directive 2003/87/EC to enhance cost-effective emission reductions and low-carbon investments, and Decision (EU) 2015/1814	2018	European Union	IMO & ICAO
IMO's challenges on the route to decarbonising international shipping: Key Issues at Stake at the 72nd Session of the IMO Marine Environment Protection Committee (MEPC 72). Briefing	2018	European Parliament	IMO
Resolution of 3 July 2018 on Climate Diplomacy	2018	European Parliament	IMO & ICAO
Council Conclusions on Climate Diplomacy	2019	Council of the EU	IMO & ICAO
European Aviation Environmental Report 2019	2019	EASA	ICAO
European Green Deal	2019	European Commission	IMO & ICAO
Greenhouse gas emissions from shipping: waiting for concrete progress at IMO level. Briefing	2020	European Parliament	IMO
Sustainable and Smart Mobility Strategy – putting European Transport on Track for future, COM(2020) 789 final	2020	European Commission	IMO & ICAO
2030 climate target plan: Extension of European Emission Trading System (ETS) to transport emissions	2021	European Parliament	IMO & ICAO
Fit for 55	2021	European Commission	IMO & ICAO
Regulation: European Climate Law	2021	European Union	IMO & ICAO

Sources Council of the European Union (2016, 2018, 2019); European Commission (2016, 2019, 2020, 2021); European Parliament (2018a, 2018b, 2020, 2021); European Union Aviation Safety Agency (2019); Official Journal of the European Union (2018, 2021)

The documents presented at the Table 1 above consist of a clear declaration of the EU towards action in mitigating GHG emissions by ships and aircrafts. In view of the adoption of legally binding measures by the IMO to reduce GHG emissions and implement the Initial Strategy, the EU appears prepared to enforce schemes that would contribute to the combating of climate change, stressing both the need for the action to be taken at the international level and the certainty that implementable systems can exist. Accordingly, after the adoption of the Long-term Aspirational Goal by the ICAO in order for the CORSIA to become an efficient scheme in decreasing the levels of GHG emissions by aircrafts (ICAO, 2022), the EU once more concocts the (re)inclusion of international aviation in the EU ETS along with a series of other measures. All this action keeps on track the two IOs who are not in favor of regional and/or unilateral measures by arguing that imbalances in the global maritime and aviation activity will appear (Van Leeuwen, 2015, 2019; Interviews no. 15, 20, 28, 34). Therefore, the EU's 'on paper' action seems to trigger procedures in both the IMO and the ICAO, or—at least—it stays committed to the goal of tackling climate change.

It is both important and very interesting to note that previous research has shown a correlation between the time of publication of the initiatives of the EU and the action taken by the two IOs (Dikaios & Blavoukos, 2023; Martinez Romera, 2018). Indicatively, two examples follow, one for each organization. In 2013, the EU published a Communication titled *Integrating Maritime Transport Emission in the EU's Greenhouse Gas Reduction Policies* (European Commission, 2013), which is the foundation of the MRV Regulation that was adopted two years later, in 2015. At the international level, in 2014, the IMO 'agreed, in principle, to develop a data collection system from ships' (IMO, 2018a), which came into force in 2018 and started being implemented in 2019. It is notable that the MRV Regulation includes a so-called *sunset clause* which foresees the development of an international system and the respective adaptation of the EU's scheme 'in order to ensure alignment with [it]' (MRV Regulation, Article 22). In ICAO, the development of the CORSIA happened because of the inclusion of international aviation in the EU ETS. Martinez Romera puts it eloquently: 'This regional measure, which has international consequences [...] features prominently in the development of the field since 2008, shaping the discussion in other realms. The inclusion of aviation in the EU ETS has pushed the envelope of international law, politics and diplomacy' (Martinez Romera, 2018: 13).

Within this context, it is noted that the Council Conclusions play a significant role in tracing the policy direction that the EU as whole should follow. For example, although the Council Conclusions on climate diplomacy are regularly published, it was only in 2009 (just before the Copenhagen Summit) that a direct mention to the IMO and the ICAO was made (Council of the European Union, 2009). In this, it was stated that the two IOs need to control the emissions from the activities falling within their jurisdiction and that this should be depicted in the agreement that will emerge through the Summit; an agreement that never saw the light of the day. Since then, mentions of the IMO and the ICAO have been made in the respective Council Conclusions of 2016, 2018 and 2019, years that could be described as milestones for the two IOs (Council of the European Union, 2016, 2018, 2019). In comparison, firm affirmations of the role that the EU should assume in order to promote more stringent measures to reduce GHG emissions and to perform an active and effective climate diplomacy are highlighted in the Council Conclusions of 2011, 2013, 2017, 2020 and 2021 (Council of the European Union, 2011, 2013, 2017, 2020a, 2021c).

4 Overt Diffusion and Procedural Diffusion: Lobbying on the Spot

Two apparatuses that play a significant part in the processes under review are those of *procedural diffusion* and *overt diffusion*. Overt diffusion engulfs the amalgamate of the vigorous activity of the EU, through its physical presence within the two Organizations, while procedural diffusion refers to the formal institutionalization between the EU and third parties. As demonstrated in Chapter 1, the EU only holds an observer status to the ICAO, while a similar scheme is in place in the IMO, with the Commission having the same status in the Organization; as such, it is clear that the relevant institutionalization in accordance with the precepts of the NPE approach is fulfilled in principle. Further, through the diplomatic action of the EU *as a whole* in both Organizations, the institutionalization of their relations is further strengthened, leading to the broadening of the activities that the European Commission, as the EU's main actor, is able to undertake within their respective frameworks. In the case studies of the IMO and the ICAO and their climate relationship with the EU, procedural and overt diffusion (somewhat) overlap, with the latter being the decisive mechanism in their association.

In order to set the background for this part of the analysis, it is interesting to note how, in her interview, Commissioner Bulc depicted the environment within which she had to operate. For her, the EU is in a slightly better position in the ICAO than it is in the IMO. In the ICAO, the EU is present at the negotiations table as an Observer, albeit not being an active member; as she puts it '[through this relation] they know us and somehow our voice is stronger officially than it is in the IMO. In the IMO we cannot apply the same strategy'. In the IMO, the Commission does not have the same status and it is expected to work more through its Ambassador and the EU Member States (Interview no. 48). These observations made by the Commissioner can also be attributed to the internal EU institutional framework, which is more advanced in the case of aviation.

In the following subsections, the specifics of overt diffusion and procedural diffusion will be analyzed for each IO, followed by some recent findings about the effect that the COVID-19 pandemic has had on the two strands of diffusion. The analysis will also have references from the wake of the Initial Strategy and the CORSIA, respectively, in order to show the incremental diplomatic action of the EU in the IMO and the ICAO, as well as the weight the Union puts in prioritizing the fighting of climate change, as per in the action taken by these two IOs. The focus will be on the period after the adoption of the Initial IMO Strategy and the CORSIA up until 2021 inclusive, to showcase the path to the adoption of specific and targeted measures and goals for the reduction of CO_2 emissions from ships and aircrafts.

IMO

The major actors that, by definition, are charged with externalizing EU's climate diplomacy in the IMO are the EU Member States. Nonetheless, the presence of the European Commission in the IMO cannot be ignored, despite the fact that it does not put the Commission in the same level as the Member States (Interviews no. 29, 30, 32). Besides the monitoring of behavior of the EU Member States in the IMO (a task mentioned both by EU officials and EU Member States officials), the Permanent EU Representation to the IMO also delivers at least two other significant functions: first, it remains in a continuous dialogue with IMO Member States, collecting information and recording interests; and, second, it extends the power of EU officials to further influence the IMO Member States.

More specifically, during the negotiations of the Initial Strategy in April 2018, in addition to the Permanent EU representative, Commissioner Bulc and a delegation of Members of the European Parliament (MEPs) traveled to London to hold meetings with less climate-ambitious IMO Member States (Interviews no. 3, 14, 15). The MEPs sounded a note of warning to the—non-EU—IMO members that, in the case of a bad agreement, or—even worse—the lack of one, the European Parliament would move forward with unilateral measures regarding GHG emissions reductions from the shipping industry (Whiteman, 2018; Interviews no. 15, 19, 20).

Furthermore, a few of the interviewees claimed that the Commission has two extra functions in the IMO, relating to the EU Member States' activity: first, it extensively coordinates on the spot (Interviews no. 3, 15, 16) and, second, in the direction of monitoring, it makes sure that the EU Member States will stick on what was agreed on the coordinating meetings (Interview no. 3). Anecdotally, an interviewee narrated an incident that took place during an MEPC meeting prior to the adoption of the Initial Strategy. A Commission official went and sat with the delegation of the EU Member State that held the Presidency of the Council of the EU at the time, checking in a way that the national delegator would advocate the Union's positions; 'like a schoolteacher, like a mother, they watch you and they write the positions for you' (Interview no. 25). This is especially apparent when the Member State holding the Presidency does not have experience or interest in the IMO issues; the Commission's stance, therefore, depends on which Member State is the chair of the Presidency and whether it is competent and involved enough to influence the dynamics of the European institutional constellation within the IMO (Interviews no. 25, 32).

No matter the slight differences in view of the processes observed by the interviewees, there is a widespread acceptance among EU Member States representatives about the influence the EU Observer has (either directly or indirectly) on the climate negotiations in the MEPCs. What differs between the EU Member States officials is the perception of the level of influence the Commission's Representative has on the IMO decisions, with some of them arguing for 'huge impacts' (Interviews no. 25, 27, 31, 52) and others being more moderate (Interviews no. 29, 32, 39, 50). There were also a few voices that categorically refused to accept that the Observer has any influence or that implied that their role should not be highlighted as much (Interviews no. 30, 46), but this belief does not seem to be confirmed by the majority of views. In contrast, it seems that the active role of the Observer seems to be getting incrementally stronger, especially after the publication of the 'Fit for 55 package', which promotes essential changes for the maritime sector that will also influence international shipping activity that entails European ports and, gives greater competence to the Commission to act both internally and externally on these subjects (Interviews no. 25, 27, 50). In parallel, the adoption of the Initial Strategy boosted the engagement of the Commission in climate shipping policies, since through this, short-, medium-, and long-term measures should be agreed on in the MEPC meetings, something that would define the exact level of ambition for each phase, and the means of implementing the Strategy.

It has to be noted that, since the adoption of the Initial Strategy, the positive environment in favor of climate agreements and targets has been substantially transformed. Although the change in the USA Administration and the much more climate-friendly proclamations made by President Biden[5] symbolize a more supportive milieu for climate change action internationally (Interviews no. 32, 46; Newburger, 2020), significant opposition has surfaced from other countries that do not wish to move forward with implementing any ambitious measures. For example, Latin American countries, with Argentina being the leading voice, Brazil, China, Russia, India, Saudi Arabia and other developing countries have a (very) negative outlook on approving short-term measures, drawing a clear divide on every aspect of the GHG emissions subject between the USA & EU axis and themselves (Interviews no. 43, 46, 52).

As during the MEPC 76 in June 2021 no agreement on the short-term measures could be seen rising in the horizon, the EU Observer stated publicly, according to an interviewee, that 'the IMO has showed us that it reached its limits, and we are terribly disappointed by the IMO capacity, and it is time for the EU to take regional measures' (Interview no. 25). This came as a surprise, since usually the EU Observer delegation keeps a low profile and its members are seen as 'the backroom boys' (Interviews no. 29, 52), as mentioning the possibility of imposing regional measures is an *issue non grata* for the IMO and other IMO Member States. Despite

[5] An interviewee called President Biden 'greener than the greens' (Interview no. 46).

this rare occasion of strong language being used by the EU, the latter claims as well that it prefers an international solution, a viewpoint that was also supported emphatically by Bulc (Interview no. 48). Regardless, some national representatives cast doubt on that, suggesting that after the struggles the EU faced and the effort it has put in developing a climate policy for shipping, it would be unlikely that it would alter its policy to such extent (Interview no. 39, 49, 50). Additionally, for some, the reference to regional measures functions as a veiled ultimatum towards the IMO, who once again, emerges as slow, very bureaucratic and 'old' organization (Interviews no. 39, 40, 49, 52).

Deriving from the above, overt diffusion plays a dual role in exporting an EU norm. Representatives of European institutions are in constant deliberation with third parties involved, promoting the EU goals with their presence, and are securing the coherence of the European position. Overt diffusion also happens through the physical presence of the EU Member States officials who advocate the EU's positions. It goes without saying, that the EU Member States carry the heaviest part of the negotiations, as they are the ones which will finally take the decisions and can officially take the floor. Because of this institutional constellation, there are voices claiming that only the Council Presidency should speak for the EU (Interview no. 25, 30). However, this may prove problematic in the peculiar situation where EU Member States whose priorities do not include maritime affairs and usually do not take part in the MEPCs (unless the issue at stake is crucial for the EU interests and demands the presence of everyone), assume the Presidency. In these cases, what happens is that they are seen 'popping up at the table' negotiating positions agreed back in Brussels (Interview no. 25, 46).

In any case, and apart from all the tactics the EU institutions employ at the international level to ensure that the EU Member States will follow the EU line, the influence of the EU as a block at the IMO is considered significant (Interviews no. 29, 31, 39, 46, 49, 50, 52). This is a result of numbers, as the EU remains the biggest coherent group within the Organization, and its voice is multiplied and amplified by the 27 (28 before Brexit) different voices of its Member States. These voices promote the goal of the EU as a whole to gain more influence and to produce more results towards a global scheme that fulfils the EU ambition to a sufficient level. As it can be expected, to this end, the primary strategy followed is outreach. Constant and deep conversations and exchanges of opinions take place between the EU Member States and the delegations of the rest of the IMO Member States, with the general attitude and atmosphere (usually coming from the northern European states) being assertive and pushy. Outreach operated by the EU Member States towards third states is also firmly encouraged from the Commission.

At the same time, in order to further advance its standing within the IMO, the EU builds its (opportunistic or steadier) coalitions, which make its positions more likely to go through, especially with small island states that are endangered because of the sea level rise due to climate change (Interviews no. 32, 46, 49, 52). Supporting this, a recent study shows how the EU engaged drastically in the High Ambition Coalition for Shipping since 2015, which up until today allows for the configuration of ambitious targets and solutions (Earsom & Delreux, 2021).[6] Another strategy that the EU has recently explored is the attempt to co-sponsor submissions with third states to be discussed during the MEPCs. Although the internal EU processes do not allow for other parties to get involved, discussions on how common submissions with non-EU Member States are going to be feasible currently take place, as combined submissions will carry more weight and will result in stronger positions (Interview no. 43).

Submissions also appear as another window of opportunity for the Commission to underline its aspiration for a dominating role over the EU Member States. When about to table a submission, the Commission finds a willing Member State and poses a request to it to prepare this submission on behalf of the EU as a whole. As the EU does not have an observer status in the IMO (only the Commission does), the latter prepares and circulates a position which starts with the phrase *'submitted by the European Commission and the EU Member States'*. This attempt to overtake the proposal is met with recoil from the Member States every single time, with the latter interfering and changing the contested phrase to *'submitted by Austria, Belgium, Bulgaria,* etc. *and the European Commission'* (Interviews no. 13, 25). This dispute also led the Commission to appeal to the

[6] It is interesting to note that in the High Ambition Coalition for Shipping the EU Member States that did not participate were Cyprus, Greece and Malta (Earsom & Delreux, 2021).

CJEU and ask for the clarification of the institutional appearance of the EU in the IMO.[7]

What emerges through the overt diffusion mostly, but also—in some cases—through procedural diffusion as well, is that the EU and the IMO speak a different language. On the one hand, the IMO speaks about what is implementable on a global scale through conditions of a consensus. On the other hand, the EU raises the bar based on a political ambition, blaming the IMO for using ludicrous arguments that do not fit the current global sentiment (Interviews no. 40, 49, 52, 53). The goal of the EU is also to make 'others' see the bigger picture of the problem and to convince them that the shipping sector must contribute at least its fair share to the international efforts against climate change. Two opposed tactics that the EU applies are spotted through the interviews (Interviews no. 25, 32, 49, 50, 52): first, the promotion of cooperation and big coalitions, as some of the third countries would 'not care less' of what the EU does on its own; and, second, the use of threats of developing a regional (EU) system which will fragment the international establishment, an advancement that the IMO despises.

It becomes evident that the EU utilizes a series of strategies, primarily influenced by the Commission's and other ambitious (mostly northern) EU Member States' demands for change. The EU as a whole arrives at the negotiations with a set agenda and very well prepared, in contrast to the rest that are not coordinated, nor they assign adequate personnel to deal with the issues on the table. It is moreover the mindset that the EU enters the negotiations with—which aims to convince its interlocutors, while the latter appear without knowing exactly what they want—that makes a big difference: 'The EU attacks and the rest defend. Always when you attack you have the initiative. It is like a bullet; the farthest it gets, the more its effectiveness reduces, but in the end, it hits the target' (Interview no. 49). In other words, the EU enters the negotiations with an ideal outcome in its bargaining quiver, which through the negotiations is mangled—at least according to several EU officials—but, eventually, an outcome towards the EU's position is largely achieved.

[7] The submission that triggered this process concerned low-carbon, zero-carbon fuels and the life cycle of GHG emissions (see for the phrasing applied, Council of the European Union, 2020b).

ICAO

As in the IMO, in the ICAO, the major agents for the EU climate norm externalization are the EU Member States. The institutional setting in the ICAO, though, is different than the one in the IMO, where all the EU Member States participate in the MEPCs. In the ICAO, the decisions for the way forward in all subjects lay on the work done by the ICAO Council within its three-year mandate. As demonstrated in Chapter 1, the ICAO Council consists of 36 States and eight (today seven)[8] out of them are EU Member States. Evidently, the power of the EU Member States when acting unified is expected to be much stronger than what it would be if it was only one actor, with one single vote (Interviews no. 38, 41, 42, 44, 47). However, not everyone agrees with this view, or at least some pinpoint challenges that the EU Member States face during the ICAO Council bargain. For them, the power of the EU is not definite, as the positions proposed by the EU Member States are sometimes controversial and challenging (Interviews no. 35, 36), as they demand a lot of effort from the EU-block's side for a certain position to prevail. *Pro rata*, this applies also to the ICAO Assembly, where 193 States participate, but the EU Member States remain an influential block. Nonetheless, the influence of the block is also boosted by the enlarged European scheme, namely the ECAC, that usually goes to the ICAO Assembly with common positions or at least mutual understandings on the issues at stake.

The EU–ICAO institutional relations run deeper than the EU–IMO ones, both in general, as the Permanent Representation is in the name of the European Union as a whole, whereas in the IMO the Permanent Representative is of the European Commission, and *specifically* in climate change issues (Interviews no. 48, 54). Concerning the role of the EU Observer, the viewpoints are mixed, but every interviewee agreed—even the representatives of the European institutions—that the current situation is adequately beneficial for the EU interests with no change needed in the institutional constellation (Interviews no. 36, 41, 42, 47). More specifically, the understanding of the power that the EU has by having 7 (or 8) votes in the ICAO Council, satisfies both the EU Member States that prefer to keep their overarching role, as well as the Commission

[8] The United Kingdom is the eighth Member at the ICAO Council. As the time framework covers to a large extent the period which the United Kingdom was an EU Member State as well, number eight will be used.

(Interviews no. 42, 47).[9] However, the case at times differs, as in certain circumstances (i.e., in climate negotiations) the Commission appears to be in deadlock (Doussis, 2014: 117–122). Thus, EU Member States officials believe that the Commission is in a very difficult position at the international level. In the cases that subjects pop up that only the EU Member States can negotiate for, the Commission Representative does not have the flexibility (or the time to get a new mandate) to negotiate and, therefore, the EU Member States ask for more flexibility so as to be able to act without European weights (Interviews no. 41, 45).

As, though, the competences of the Commission in aviation matters are well-defined and well-based in the internal EU coordination processes, it is interesting to examine the arrangements between the EU Representative in the ICAO and the EU Member States. Similarly to what takes place in the IMO, there is again the two-level function of the Union within the ICAO: first, the on-the-spot coordination between the EU stakeholders and, second, the similarly on-the-spot lobbying among the EU as a whole and the rest of the ICAO Members and institutions. It has to be noted again that the EU has an enhanced role as an Observer in the ICAO, which is also established through various memoranda of cooperation (e.g., see Official Journal of the European Union, 2011). This ostensibly gives a more prominent role to the EU Delegation in the ICAO, with some EU Member States officials recognizing its significant role and influence (Interviews no. 38, 44), while others rejecting its impact or minimizing its presence to an intra-EU managerial nature (Interviews no. 34, 41, 45, 51).

It is true that, through the interviews, the role of the EU Representative seems to be focusing more on the intra-EU relations, rather than the EU–ICAO relations, at least on climate change (Interviews no. 34, 41, 54). The main responsibility of the EU Representative seems to be to coordinate all the EU Member States and especially the ones which participate in the CAEP. Weekly coordination meetings take place, during which the EU position is reminded, as well as what has been discussed in the ICAO Council. Thus, the EU Representative is in constant contact with Brussels and watches closely what is going on in Montreal. EU-wise,

[9] It is interesting to note that officials who have participated in the negotiations in the past, argued in favor of a single EU membership in the ICAO, as the EU Member States were acting more autonomously than anticipated by the European institutions (Interviews no. 35, 40).

the Representative is in continuous cooperation with a European coordinator for aviation matters coming from the Presidency of the European Council, unwrapping a kind of dual coordination process that takes place on the one hand between the EU Member States alone and on the other hand with the Union as a whole. At the same time, the weekly coordination meetings also include countries that are not EU Member States but are simultaneously ECAC Members and ICAO Council Members. The coordination with these countries allows for the aforementioned creation of broader coalitions. The same happens for environmental and climate topics, with countries that participate in the EAEG. Through this cooperation, an agreement on the line that will be defended in the ICAO Council is reached. It goes without saying that these repetitive three-level coordination meetings allow for the European (region-wise) voice to be heard louder and to have better results on the aspired European/EU-path. Besides, there is a general agreement on how much impact the EU as a whole has on ICAO decisions and decision-making regarding climate change issues (*see Fig.* 3, *Chapter 5*), and non-EU countries who wish to join the room with the block, are more than happy to follow the norm(!).

The tactics that the EU employs in order to be influential in the ICAO are of particular interest, especially regarding how the diplomatic apparatus of the Union functions and how much it matures over time. It is accepted that the EU Representative has an instrumental role and is mandated to coordinate the EU Member States, which will bear the diplomatic burden of influencing the ICAO. This situation, at times, poses challenges to the EU as a whole, as opposition emerges from the rest of the ICAO Member States. The most common complaint regards the 'single voice mantra' of the EU, which has led third states to express the viewpoint that if the EU Member States support the exact same position, they should acquire only one voice and one vote in the ICAO.[10] Thus, there has been an 'unofficial' agreement and guideline to negotiate the same position with different words or to distribute the burden sharing between different negotiating points; this, though, happens in an organized way (Interviews no. 28, 34, 36, 38, 41, 42, 47, 51).

[10] One interviewee expressed the concern that the Commission is trying to be the representative of the EU Member States, which would also lead to 'one voice' and would enhance the argument by third states in favor of 'one vote' (Interview no. 36). This is not confirmed by the Commission officials, who claimed that the current situation is beneficial for both the Commission and the EU Member States (Interviews no. 42, 47).

The above situation is much different from what was taking place in the ICAO before 2016. Between 2008 and 2013, the relations among the EU and the ICAO experienced great turbulence due to the inclusion of international flights in the EU ETS, which eased between 2013 and 2016, as the EU realized that it needed to approach the ICAO in different terms and has been smoothed out from 2016 hitherto. Without going into much detail, it is noteworthy that during the period of the friction (2008–2013), EU delegates were seen as the 'black sheep' of international aviation with boos being under way every time they were taking the floor, especially during the ICAO Assembly in 2013 (Interview no. 35, 47, 53). As said earlier, the EU was compelled to initiate the 'Stop the Clock' regulations, and it took about a decade for the EU to be seen again as a trustworthy interlocutor. During the 2008–2016 period, it was not only the EU Member States and the European Commission that were seen as rivals of free international aviation. They were joined also by the European Parliament, which is always much more polemic in favor of stringent climate targets to be adopted by civil aviation. The Committee on the Environment, Public Health and Food Safety (also known as the ENVI Committee) of the European Parliament, which is responsible, among others, for the scrutinization of the EU institutions' actions regarding the issues concerning climate change, faced even greater backlash that was topped with the refusal of the ICAO to allow the attendance of an MEP delegation in the CAEP proceedings in 2016. Neglecting the refusal, the MEPs traveled to Montreal and after diplomatic efforts managed to enter gradually, first in social events, then in the CAEP and finally to the Assembly, in the negotiations (Interviews no. 35, 48, 53). The anecdotal storytelling is aimed to showcase how much the European Parliament wanted to pressure for measures to tackle climate change from civil aviation activities. By being present during the international bargaining for these targets, it insinuated that—as in the IMO later on—if the ICAO did not go big, the EU would once more strive for ambitious regional action with global consequences—adding to the narrative that without the EU nothing would have happened regarding the mitigation of GHG emissions in ICAO.

This narrative remains very strong up until today (Interviews no. 28, 38, 41, 42, 44, 45, 47, 54), without though neglecting the fact that the EU was not the decisive actor for the CORSIA to be adopted. It is a common understanding between the EU Member States officials that the EU as a whole acts more as a broker, a bridge-builder and an influencer, but not as a deciding group on its own (Interview no. 36). The recognition of the impact of the EU ETS on how CORSIA was build is widespread (Interviews no. 36, 40, 44, 47), but also the huge influence of the industry (Interviews no. 36, 48) is highlighted. In contrast, after the adoption of the CORSIA, as the international sentiment in favor of climate targets diminished, the EU stands alone in the fight against climate change within the ICAO (and not only). In fact, the presence of the EU in the ICAO is of great importance as it proves that CORSIA can be implemented and thus no stepping down or going back can be argued by the rest of the countries (Interview no. 28).

The 2019–2022 mandate for the CAEP revolves around the feasibility of an LTAG, which will lead the anticipated successful implementation of CORSIA, through elaborating the technical aspects and aviation fueling. In this process, the role of the EU as a whole is deemed as instrumental, and the officials interviewed feel that the EU is the only pioneer in the field of climate change and that it needs to keep pushing for collective action (Interviews no. 34, 41). Moreover, they are conscious of the influence that the EU has on the ICAO, in spite of the small (as seen by several) steps taken, that are, though, in the right direction. 'The EU is strong enough to still put climate change as a top priority' (Interview no. 37) and is exceedingly active in order to achieve its goals (Interviews no. 28, 33, 44, 47).

What needs to be mentioned is that in order to be successful in the above processes, the EU should not be anchored to the Brussels bubble and forget or disregard global politics and the fact that aviation is really international. Unlike the more 'threat-based' approach that is seen within the IMO framework, in the ICAO—maybe because of the previous failure—there is an understanding that dictates that no other country can be blackmailed to do something for climate change; the action needs to always come through collaboration. As the ICAO moved towards the adoption of the LTAG, there is a feeling that the ICAO Secretariat moves closer to the Union's positions. This might be the result of the technical expertise that the EU generously shares with the ICAO,[11] and/or because the EU learned how to never isolate during the negotiations but to keep open communication channels with third states and to be

[11] For more see next Sect. 5.

open to incorporate (some of) their concerns in the discussions (Interviews no. 42, 47). In this process, the EU Representative in Montreal is admittedly particularly active and their role is crucial (Interviews no. 37, 41, 44). The EU Representative is in constant deliberations with third countries, speaks with the big actors (something that small EU Member States would not have been able to) and keeps track of whatever happens within the ICAO, and is there offering to transfer the necessary knowledge. Hence, the EU Delegation enhances its position within the ICAO as coordinator of the EU Member States but also of third states, through the development of coalitions and the power of the gathered information and, all in all, knowledge and expertise.

Wrapping up, the Union seems to believe that it has a significant amount of influence on the ICAO proceedings. This is largely due to the fact that the EU, especially during the last few years, performs as a solid block but appears conciliatory in its relations with third parties. 'No EU Member State has diverged for many years now; if the system remains the same and the EU continues to know how it works and how to approach it, it is going to serve our interests' (Interview no. 47).

Finally, one more noteworthy difference, compared to the IMO, is spotted through the interviews; it is the admission that the EU Member States officials in the ICAO are highly experienced, and their delegations have very good personnel. This is probably the result of the longer period that officials participate in the ICAO, making them themselves much more 'institutionalized', while in the IMO the turnover of officials happens much more often—as it appears from the official participation documents of both Organizations (2019a, 2018c, 2019, 2021; ICAO, 2013, 2016; IMO, 2018b). The reason for this difference may be that the CAEP is a smaller institutional structure than the MEPC and the links and bonds that are created within the former strengthen, at the same time, the understanding with which the issues should be approached. Another aspect which needs to be mentioned concerning the diplomatic personnel is that in the CAEP aviation people with environmental expertise participate (while remaining national representatives), whereas in the ICAO the Council and the Assembly are composed by national diplomats who do not have the technical expertise to take this role. Therefore, there have been observations where in CAEP (or even before an issue emerges in the ICAO framework), experts have already come to an agreement, although the situation might sometimes become much more complicated at the political level, because some Member States may hold a different position (Interviews no. 36, 41). This also marks a difference with the IMO, where, nowadays more and more, diplomats who have participated in the UNFCCC negotiations travel to London and take part in the process in order to make the climate change subject political and to achieve ambitious goals, rather than letting it to the technical experts coming from different Ministries (Interview no. 41).

COVID-19 Effects on (Climate) Diplomacy

The spreading of the COVID-19 pandemic introduced drastic changes in the diplomatic arena, as it did in every other aspect of everyday life (Maurer & Wright, 2020). Thus, its impact on climate diplomacy has been greater in the field of overt and procedural diffusion than any other. The IMO and the ICAO are no exception to the impact of the restrictions imposed due to COVID-19, with all regular meetings moving to a virtual format (some of them had even being cancelled, especially during the first quarter of 2020). Divergences in the standpoints of the participating officials on how this new reality impacted the procedures can be found in both IOs.

In the IMO, the point of view that the meetings were significantly obstructed prevails (Interviews no. 25, 32, 39, 43, 46, 52), although a couple of interviewees claimed that the influence was minor and the process was not slowed down (Interviews no. 27, 29). Contrary to the latter opinion, all the rest officials argued against the online meetings that were significantly shorter in time—only three hours per day in contrast to five all-day in person meetings—and that the majority of the participants connected with fixed positions and no leeway and space to maneuver, reading only written formal statements and unable to perform any flexibility. There were also complaints about the lack of the activity taking place outside the plenaries, such as corridor talks, coffee breaks and canteen meetings. This activity builds trustworthy relationships between the negotiators and facilitates important decisions. Apart from the constraints that the physical absence imposed, virtual meetings also have had negative repercussions for the actual targets. According to the interviewees, the online meetings lowered the productivity of the negotiations and watered down the overall ambition; only 1/3 of the work that is usually done was proceeded and significant delays in achieving the anticipated outcomes of the Initial Strategy were noted.[12]

In the ICAO, the image is different. The EU Member States officials are split in two equal groups on the subject. The first group supports that nothing changed during the pandemic, the work continued normally and 'everything happened as it was supposed to happen' (Interviews no. 34, 37, 44). The second group claimed that not meeting your interlocutors in person made it very hard to negotiate and find consensus, a situation that significantly confined the ambitions, allowed for several reservations to surface (especially for the LTAG) and led some positions to be really 'stuck' with no solutions able to be found (Interviews no. 41, 45, 51).

As it was expected, the one-to-one networking and socialization were missed and described as very much needed for both groups. The Commission officials lined up with the second group's arguments and even claimed that it is better for the General Assembly scheduled for Fall 2022 not to take place at all if it is to happen virtually. For them, the limit of the online meetings has been reached, as the progress made is far from ideal (Interviews no. 42, 47, 54). As put by one national diplomat, 'even if the EU Member States might have some reservations, there is a common European policy in place, and it will not be blown up by COVID-19' (Interview no. 41).

5 Transference: Investing in Technical Development

According to Manners (2002: 245), diffusion through transference 'takes place when the EU exchanges [...] technical assistance with third parties through largely substantive or financial means'. What Manners fails to acknowledge in this regard is that diffusion is an extrovert activity which does not, by definition, entail exchange. The exchange might happen one way or another, but it is not a crucial factor when the diplomatic action of one entity towards another is examined, especially through diffusion; replacing the word 'exchange' with 'invest' proves more useful.

The EU recognizes that by providing its expertise or economic assistance to countries that need them, it will be more successful in its

[12] A fun fact regarding this process was the role that Argentina played, as it obstructed the meetings by filibustering; this resulted to its removal from the IMO Council in the elections of December 2021.

endeavors. Commissioner Bulc stressed the important role of this action, by also putting it in the greater context of a *just* move, which will support states lacking in knowledge and resources to develop the necessary capabilities in order to implement international policies that will benefit the international society.[13]

In the IMO, a few years before the adoption of the Initial Strategy and in parallel with the preparations of the MRV Regulation in 2015, the EU invested ten million euros in a project titled *Capacity Building for Climate Mitigation in the Maritime Shipping Industry* to run for four years (2016–2019). This project was implemented by (and through) the IMO and created five centers of excellence around the globe, aiming to enhance technical cooperation, capacity building and technology transfer in regard to GHG emissions reductions from shipping activities. According to a couple of interviewees, this project helped to curb doubts, especially from developing countries, about the Initial Strategy, extending thus 'a kind of influence' (Interviews no. 15, 22, 39, 52). The project 'was very well received' (Interview no. 3) and was expected to continue but, unfortunately for the EU interests and according to a Commission's officer, 'we have not been able to continue supporting this project for now -due to the reconstruction of the DG for Developmental Aid- and we currently need a levy for funding that will help us' (Interview no. 52).

In a more unofficial manner that was mentioned during the interviews, the EU seems to also finance or give additional assistance to small and/or vulnerable states to participate in the IMO. Moreover, the EU prepares for (instead of) them their paperwork and supports their positions with technical knowledge (Interviews no. 46, 49).

The idea for financing a project in the IMO came from a similar project in the ICAO, titled *Capacity Building for CO_2 Mitigation from International Aviation*, that had been funded by the EU in the late 2013. The EU had invested six and a half million euros and the scope of the project was to assist fourteen countries in Africa and the Caribbean to adapt to the new climate reality of aviation in a period of five years. The project could be characterized as a moderate success, as half of the beneficiary countries entered CORSIA on a voluntary basis from the beginning of its implementation period (ICAO, 2019). Through the two capacity

[13] See quote in page vi.

building programs, the EU's goal was to transfer its know-how to achieve the diffusion of its climate norm globally.

For the interviewees, the financial contribution to the ICAO for capability and capacity building is important, as—this way—the EU first eased the concerns of third parties for the costs (Interview no. 51) and, second, through budgetary policies 'you can make yourself heard' (Interview no. 28, 37). 'Financing capacity building is not always linked with the EU's influence but quite few times it is. Third states are more open to work with the EU under CORSIA, when for example they get trade incentives' (Interview no. 42). More important, though, seems to be the transfer of technical expertise and the standards setting. For the latter, 'the EU is more catholic than the Pope' (Interview no. 47), with the ICAO Secretariat being open and receptive to take the know-how the EU produces and to try to find global solutions based on it, leading thus to a de facto influence of substantial normative characteristics (Interview no. 36, 38, 42).

6 Cultural Filter: Inferring and Integrating Norm Diffusion

According to Manners, cultural filter 'affects the impact of international norms and political learning in third states and organizations leading to learning, adaptation or rejection of norms' (Manners, 2002: 245). Having established that the EU has one of the most advanced climate policies on GHG emissions reductions and that it wishes to export its climate norm in order to have a common understanding with third parties on how climate change should be tackled, the functions of learning and adaptation—or sometimes rejection—are a logical consequence on what happens when the EU interacts with third parties.

Learning and adaptation takes place throughout the institutional interaction of the organizations under examination, with the aim to implicitly lead to a change of what is considered 'normal' in a specific point in time. More specifically, for both the IMO and the ICAO, the EU has adopted its own initiatives and regulations that are not only meant to be implemented within its boundaries, but it aspires to have them exported at the international level. This aspiration stems unintentionally in terms of unintended consequences because of the mere existence of the EU's framework *(contagion)* and, also intentionally, in terms of strategically planned interventions that are aimed to influence the IMO and the ICAO in a certain manner *(informational diffusion)*. Moreover, the Union, through its formal relations with the two IOs at hand which perform an array of functions and assist both learning and adaptation, interacts with officials from third parties. These established relations further allow the physical (or virtual due to the COVID-19 pandemic) appearance of additional EU officials apart from the permanent EU representatives and enhance the transfer of legal and technical know-how, as well as ideas, which one way or another are constantly provided to the two IOs *(procedural and overt diffusion)*. To ensure the *transference* of its technical expertise, as mentioned earlier, the EU also invested a certain amount of funds to facilitate the acceptance of the normative alterations it strived for. Consequently, in a way, the cultural filter seems to be the underlying logic and integrating stage of norm diffusion, while—at the same time—being what allows for the normative change to take place in the end.

All the above can either result in a successful change of an actor's 'normal' or to a rejection of this process. During the previous decade, learning and adaptation—which had long been present in the institutional interplay between the EU and the two IOs—culminated in the adoption both of the Initial IMO Strategy and of the CORSIA by the ICAO (Dikaios, 2020). This was also depicted in the literature of the previous decade for both Organizations. For the IMO, Van Leeuwen and Kern comment that 'the EU is a driving force in international environmental shipping because it changed IMO decision-making and improved the implementation of IMO standards binding and enhancing their enforcement' (Van Leeuwen & Kern, 2013: 70). Accordingly, for the ICAO, the comment by Martinez Romera (2018: 13) on pushing the envelope of the international developments describes picturesquely the EU's influence. Since then, though, opposition has been growing against the EU's ambitions in both IOs in the period after the adoption of the two strategies. This opposition seems to put brakes on the anticipated changes that the EU would like to see happening in the business-as-usual processes of the IOs, leading to a path that it may end to rejection of its climate norm and to the mitigation, thus, of its normative power.

Taking this observation as a starting point, the next section evaluates the EU's normative climate diplomacy towards the IMO and the ICAO.

7 Concluding and Discussing the External Practices: Evaluating the EU's Climate Diplomacy Through Coherence and Effectiveness

The exercise of evaluating the diplomacy of the EU has not attracted significant attention in the literature, mainly due to the complex institutional framework that governs the EU's diplomatic performance (Kleistra & Van Willigen, 2015). In that sense, evaluating diplomacy *lato sensu* has been characterized by Kleistra and Van Willigen (2010) as *a mission impossible*. Nevertheless, the two authors have argued that measuring the EU's diplomatic action through a linear outcome performance, which gives prominence to the input (here, internal practices), the output (here, external practices) and the outcome (here, diplomacy evaluation) can be a sufficient method to evaluate this action when the objective of diplomacy is clear (Van Willigen & Kleistra, 2013). In our case, the objective of the climate diplomacy of the EU is crystal clear: to make the IMO and the ICAO to adopt effective tools in order to mitigate GHG emissions from ships and airplanes, respectively. Thus, the evaluation quest of the EU's diplomacy reaffirms the claim that the internal cannot be separated from the external. Having analyzed and assessed the internal processes, what remains to be examined is, first, whether the EU performs diplomatic cohesion and, second, whether the employment of the mechanisms offered by the NPE approach sufficiently serve the EU (as a whole) goals.

Diplomatic cohesion, i.e., the expression of the common negotiating position which was agreed back in Brussels from all European stakeholders during the international negotiations, is confirmed through the analysis. Despite the voices which claim that more flexibility at the international level would be preferable (Interviews no. 25, 29, 38, 45), in the end, the same, single, EU position is advocated.[14] While in the IMO the homogeneity of the EU voice seems to be a prerequisite in the Commission's strategy for the EU position to appear as strong as

[14] No mention to voting cohesion is made here, as the vast majority of the negotiations at the two IOs conclude in agreements through consensus. In the cases a vote was needed, even EU Member States with no special interest on the matters of maritime or aviation, appeared in the plenaries to vote in favour of the EU's position (Interviews no. 12, 25, 46).

possible, this is not the case in the ICAO, where a certain level of heterogeneity in discourse is embedded in the negotiating practice of the EU. In other words, it emerges that, in the ICAO, the representatives of the EU Member States can support the same negotiating position by using different words or bargaining approaches, sharing, along the way, also the burden of who says what. While in the IMO burden sharing can also be found in the submissions' procedure, the requisition of the Commission is that during the negotiations no diverging voices are heard; a point that is confirmed through practice. The reason behind this difference in approaches can be located, once more, to the better-defined institutional constellation of the ways the EU works towards the ICAO, which allows for more flexibility, and the ongoing battle between the Commission and the Member States for the arrangements in the IMO, making the Commission to appear more dominant in its approach.[15] Nevertheless, and despite the negative answer by some of the interviewees to the question regarding the establishment of a single EU voice, at the same time, several of them argued that a coherent position serves for enhanced credibility of the EU as an actor; fragmentation leads to the diminishing of the negotiating power (Interviews no. 5, 7, 9, 12, 16, 17, 19, 26, 29, 31, 34, 41, 44, 51). For Commissioner Bulc, in fact, it was/is very important that everyone was/is telling the same story (Interview no. 48).[16]

Therefore, accepting as a given the fact that coherence is performed, what effect do that rest of the external practices, namely the NPE mechanisms, have? Contagion and informational diffusion set the basis and draw the intentions of the Union towards the international level; together, they show the long-term commitment of the EU in putting efforts towards convincing third actors to engage in the reduction of GHG emissions. It can be argued that both mechanisms perform satisfactory outcomes, as several international developments have taken place chronologically after the EU's actions.[17] Here, a correlation can be detected, in accordance with the admission, as noted both in the literature and in the interviews,

[15] A shift towards burden sharing in a similar manner to aviation's policy seems to be emerging in the maritime policy, with countries taking gradually the lead in different portfolios (Interview no. 50).

[16] She also included the industry and the NGOs.

[17] A couple of the interviewees praised the role of the timing that the EU/Commission decided to go public with the policy initiatives and declarations concerning climate change rules in maritime and aviation affairs (Interviews no. 5, 26).

that without the EU nothing would have happened into that direction. It is thus the advanced climate governance system of the EU, in conjunction with purposeful policy initiatives (with intended externalities) that lead the way of international progress on the matter of climate change. This, of course, does not mean that the EU is the only decisive actor in the international climate arena, but that, through the action it performs via several mechanisms aiming to combat the negative effects of climate change, positive results are achieved. In a turn of phrase, an interviewee said, 'I do not want to ignore or downplay the role of the EU, but we need to take into account that the power play is not between the USA and the EU anymore, but between the USA and a lot of other actors' (Interview no. 36). On the one hand this might be true, on the other, it confirms the significant role of the EU in putting the fight against climate change on the international agenda.

The ways the EU put the fight against climate change in the agenda of the IMO and the ICAO have been deliberated on in previous research. The two high profile strategies on climate change that have been adopted by the IOs at hand, namely the Initial Strategy and the CORSIA, have a European stamp on their inception, albeit their final format. In more detail, the Initial IMO Strategy, as stated in its title, is a strategy that does not have any legal obligation deriving from it, in spite of the (rather) ambitious goal of 50 per cent reduction of GHG emissions by 2050. Instead, it only describes what the next steps that should be taken are, in order for this goal to be achieved, deferring any mandatory decisions to the future. Respectively, based on a different rationale, the CORSIA is an offsetting scheme that does not foresee the mitigation of GHG emissions at all. Although both agreements were received with laurels, none of them were close to what the EU would like to have had concluded upon. For the IMO the negotiating position was 70–100 per cent reduction by 2050,[18] while for the ICAO a system closer to the initially anticipated EU ETS would have been much appreciated (Interviews no. 35, 40). Consequently, although the European climate norm is exerted at the international level—and to a large extent this is attributed to the EU—only the hat survived and no cattle.

[18] For some of the interviewees this was just a strategy to achieve the highest possible outcome (Interviews no. 8, 16, 20), but for others this was exactly what the Commission would have liked to see in the final output (Interviews no. 1, 14, 15, 22).

Having, thus, these strategies as a baseline, the EU through its active diplomatic performance strives for the best possible result in the post-strategies period, aiming to ensure that obligatory measures will occur and will be put in place. To this effect, during the 2023 Initial IMO Strategy revision process and the 2022 ICAO General Assembly in which the LTAG was discussed, the EU put significant efforts to convince its counterparts to engage with reducing GHG emission in a systematic and substantial manner and stop exercising the fine art of stalling or agreeing only on picking the low hanging fruits.

Nevertheless, some of the issues that arise can be attributed to the EU's diplomatic action towards the two Organizations. A first pitfall of the EU's action is the meager investments it distributes in order to curb doubts for the developing world and lead to an agreement in rules that would have an impact on their development. Commissioner Bulc said, characteristically, that she understands the difficult position these countries are in and that they have very different needs from the developed countries, and it is more than logical to provide them with resources in order to acquire sufficient expertise and, eventually, to contribute to the ultimate goal of mitigating GHG emissions (Interview no. 48). The problem, here, arises from the fact that the EU does not seem to be investing adequately to this objective, failing to uphold to one basic rule of strategic studies that 'he who gives, dominates' (Fernand Braudel at Katsoulas, 2022: 1). For both sectors, the EU funded two projects in the mid-2010s, which indeed culminated to the support of a large percentage of the EU's positions in the international negotiations. 'The EU needs to understand what other countries' priorities are; we do not have the same priorities. We need to invest in these countries and find the instruments to satisfy them if we want them to get on board' (Interview no. 45). Since then, though, no more funds have been secured for these purposes, unpacking a situation where the EU voice has again become somewhat irrelevant.

This is also confirmed by the stance that is currently being unfolded. The aggressive position expressed by the Commissioner of Transport, Adina Vălean, to 'slap' the IOs with some measures shows two things. First, that the Commissioner believes that diplomacy has failed to bear fruit at the international level and, second, that the EU is ready to return to a unilateral course of action regarding ship and aircraft emissions—in the name of making Europe a climate-neutral continent by 2050. For the policymakers/stakeholders/interviewees, this development sounds alarming as their firm belief is that only through negotiations and by (learning to) listening (listen) to the needs of third parties, progress can be achieved. Additionally, for them, the development of common positions is the only way forward if we [as the EU] are to secure a system which will contribute to the mitigation of GHG emissions (Interviews no. 28, 33, 36, 41, 43). However, as put by an interviewee, 'I am not convinced that in the EU we are open-minded and ready to listen' (Interview no. 36).

For the record, it is very interesting to make a reference to the effects that the EU's climate diplomacy, especially through its actual presence in the two IOs (overt and procedural diffusion), has had, diplomatically-wise. It is commonplace between the interviewees that the EU as a whole has a significant level of influence both in the IMO and the ICAO (Interviews no. 25, 26, 28, 34, 37, 43, 49), albeit for some not as effective as the EU would expect it to be (Interview no. 35, 39, 40). However, due to the active climate performance of the EU, rival opinions have emerged and opposition has formed within both IOs. For example, an unexpected claim that 'shocked' the EU was that, in the ICAO, the least developed countries argued in favour of the expansion of their aviation sector and connectivity regardless of the emissions and their impact (Interview no. 40).[19] Of course, opposition came and still comes from the usual suspects (Russian Federation, Brazil, India, Saudi Arabia, to name just a few) that do not care for any climate measures to be in force (Interviews no. 28, 34, 37, 41).

Yet again, the different points in time in which the EU's engagement happened in the two IOs seem to offer explanations on the different levels of opposition that have occurred. After the 'great opposition' in 2012 and the upend of the EU's position in the ICAO, the complaints about the single EU voice (especially during the 2010s) resulted to the on-purpose heterogeneity discussed earlier. On the contrary, in the IMO where the involvement of the EU as a whole is still rather fresh, opposition has emerged as an answer to the European coherence. The recoil that the EU is facing, combined with the rather indifferent and lukewarm

[19] The integration of the concerns of the developing world in any scheme developed within the ICAO and the implementation of the *Common but Differentiated Responsibility* principle, which foresees that the burden to mitigate GHG emissions falls more on the shoulders of the ones who have created the problem in the first place, has been a sticking point in the ICAO negotiations (Erling, 2017).

final outputs, i.e., the Initial IMO Strategy and the CORSIA, leave the EU's normative climate diplomacy exposed.

According to the diplomacy evaluation theory (Kleistra & Van Willigen, 2015; Van Willigen & Kleistra, 2013), the broader international framework within which the diplomatic action takes place and the results it produces, need to be taken into account. Although it falls out of the scope of this book to systematically engage with the 'others' (e.g., third countries, the industry, NGOs), the rejection of the EU's climate norm can be proved through the following: the IMO and the ICAO adopted two ostensibly ambitious strategies to combat climate change effects from the action of ships and aircrafts, respectively. Nevertheless, while this is viewed as a success of the EU's (normative) climate diplomacy towards both IOs, a closer look of the actual outputs, as well as of the developments that have taken place since then in order to define the implementation aspects of the two strategies, tells us a different story.

Further, the point of views expressed from the experts regarding the impact of the COVID-19 pandemic to the evolution of the EU's climate diplomacy leads to the conclusion that in order for the EU climate norm to be effectively exported towards the IOs, a necessary condition is the overt diffusion taking place. Thus, and taking into account the significant delays that have been characterizing the period since the adoption of the Initial IMO Strategy and the CORSIA, it is possible to argue that the normative power of the EU has diminished throughout the COVID-19 pandemic; this has happened especially due to the lack of physical interaction, which according to the theoretical model is a necessary factor. The pandemic, though, is not to be blamed (at least more than just as an extra obstacle to the whole process), since the majority of the interviewees, if not all, focused on the rough period and relations for agreeing stringent and tangible measures for mitigating GHG emissions from shipping and aviation. This is due to the hardening of the position of several countries around the world that severely object to rules that would gravely impede, or even eliminate, their trade and movement activity. The COVID-19 pandemic has functioned as a lever for their arguments, pinpointing the extreme impacts that the lockdowns had, especially on aviation, which has changed the dynamics in favor of slowing down the implementation of CORSIA and the avoidance of setting an LTAG. In the IMO, the impacts of the lockdowns were mostly used as an opportunity of (or an excuse used by) several third countries to obstruct the *greening* processes, resulting in no progress being made during the negotiations, leading to a perpetual postponement of taking action towards fighting climate change. Anecdotally, 'the new trend is asking for impact assessments for everything; reluctant (both EU and non-EU) states hide behind this constructed need, not saying bluntly that they do not want climate change action' (Interviews no. 42, 46, 53).

As a result, the two work assumptions formed to examine the external practices of the EU as a whole, i.e., *WA5: The EU is externally effective when it speaks with a single voice* and *WA6: The EU is externally effective when the NPE mechanisms apply in the case under consideration*, are both, at least for the time being, rejected. Even in cases where heterogeneity is allowed to serve the single EU voice or where coherence is performed, the results of the negotiations do not seem to satisfy the EU's interests. Accordingly, even when the NPE mechanisms apply perfectly in analyzing the EU's action towards third parties and the normative type of its diplomacy can be boldly affirmed, the effectiveness of this action is not guaranteed. Thus, although the precepts of the two work assumptions are confirmed and the EU's diplomacy functions according to the theoretical expectations that would make it successful, their essence has not shown progress and, while the EU (climate) norm is externalized, it falls on deaf ears. Neither the IMO nor the ICAO, in spite of their resolutions seem to sincerely put efforts to mitigate GHG emissions from the actions under their jurisdiction.

List of Interviews

Interview no. 1, Advocate of the industry of an EU Member State, phone call, 31 March 2019.
Interview no. 3, EU Member State official, phone call, 15 April 2019.
Interview no. 5, EU Member State official, phone call, 16 April 2019.
Interview no. 7, EU Member State official, phone call, 25 April 2019.
Interview no. 9, EU Member State official, phone call, 30 April 2019.
Interview no. 12, EU Member State official, phone call, 2 May 2019.
Interview no. 13, EU Member State official, phone call, 3 May 2019.
Interview no. 14, Advocate of the industry of an EU Member State, phone call, 7 May 2019.
Interview no. 15, European Parliament official, phone call, 7 May 2019.
Interview no. 16, EU Member State official, email exchanges, 10 May 2019.
Interview no. 17, EU Member State official, phone call, 15 May 2019.
Interview no. 19, European Commission official, phone call, 22 May 2019.
Interview no. 20, EU Member State official, phone call, 22 May 2019.

Interview no. 22, EU Member State official, phone call, 24 May 2019.
Interview no. 25, EU Member State official, virtual call, 2 November 2021.
Interview no. 26, Advocate of the industry of an EU Member State, virtual call, 10 November 2021.
Interview no. 27, EU Member State official, virtual call, 11 November 2021.
Interview no. 28, EU Member State official, virtual call, 16 November 2021.
Interview no. 29, EU Member State official, virtual call, 17 November 2021.
Interview no. 30, EU Member State official, virtual call, 17 November 2021.
Interview no. 31, EU Member State official, virtual call, 17 November 2021.
Interview no. 32, EU Member State official, virtual call, 18 November 2021.
Interview no. 33, EU Member State official, virtual call, 24 November 2021.
Interview no. 34, EU Member State official, virtual call, 26 November 2021.
Interview no. 35, European Parliament official, virtual call, 26 November 2021.
Interview no. 36, EU Member State official, virtual call, 30 November 2021.
Interview no. 37, EU Member State official, virtual call, 1 December 2021.
Interview no. 38, EU Member State official, virtual call, 1 December 2021.
Interview no. 39, EU Member State official, virtual call, 3 December 2021.
Interview no. 40, European Commission official, virtual call, 3 December 2021.
Interview no. 41, EU Member State official, virtual call, 3 December 2021.
Interview no. 42, EU Member State official, virtual call, 3 December 2021.
Interview no. 43, EU Member State official, virtual call, 6 December 2021.
Interview no. 44, EU Member State official, virtual call, 7 December 2021.
Interview no. 45, EU Member State official, virtual call, 7 December 2021.
Interview no. 46, EU Member State official, virtual call, 8 December 2021.
Interview no. 47, European Commission official, virtual call, 8 December 2021.
Interview no. 48, Violeta Bulc, European Commissioner of Transport (2014–19), virtual call, 20 December 2021.
Interview no. 49, Advocate of the industry of an EU Member State, phone call, 3 January 2022.
Interview no. 50, EU Member State officials, virtual call, 4 January 2022.
Interview no. 51, EU Member State official, virtual call, 10 January 2022.
Interview no. 52, European Commission official, virtual call, 24 January 2022.
Interview no. 53, European Parliament official, in person, 31 January 2022.
Interview no. 54, European Commission official, virtual call, 10 February 2022.

References

Council of the European Union. (2009, October 21). *Council Conclusions on EU Position for the Copenhagen Climate Conference (7–18 December 2009)*, 2968th Environment Council Meeting.
Council of the European Union. (2011, July 18). *Council Conclusions on EU Climate Diplomacy*, 3106th Foreign Affairs Council Meeting.

Council of the European Union. (2013, June 24). *Council Conclusion on EU Climate Diplomacy*, Foreign Affairs Council Meeting.

Council of the European Union. (2016, February 15). *European Climate Diplomacy after COP21—Council Conclusions (15 February 2016)*, 3447th Meeting.

Council of the European Union. (2017, March 6). *Implementing the EU Global Strategy—Strengthening Synergies between EU Climate and Energy Diplomacies and Elements for Priorities for 2017—Council Conclusions (6 March 2017)*, 3525th meeting.

Council of the European Union. (2018, February 26). Council Conclusions on Climate Diplomacy—Council Conclusions (26 February 2018), 3598th Meeting.

Council of the European Union. (2019, February 18). *Climate Diplomacy—Council Conclusions (18 February 2019)*, 3673rd Meeting.

Council of the European Union. (2020a, January 20). *Council Conclusion on Climate Diplomacy*, 3742nd Meeting.

Council of the European Union. (2020b, January 31). *IMO—Draft Union Submission to the Seventh Meeting of the Intersessional Working Group on Reduction of GHG Emissions from Ships of the IMO in London from 23 to 27 March 2020 on the Introduction of Life Cycle Guidelines to Estimate Well-To-Tank Greenhouse Gas (GHG) Emissions of Sustainable Alternative Fuels to Incentivise the Uptake of Sustainable Alternative Fuels at Global Level—Endorsement*, 5606/20.

Council of the European Union. (2021a, May 11). *76th Session of the IMO Marine Environment Protection Committee (10–17 June 2021)—EU Coordination*. 8277/1/21 REV 1 LIMITE.

Council of the European Union. (2021b, November 15). *77th Session of the IMO Marine Environment Protection Committee (22–26 November 2021)—EU Coordination*. 13527/1/21 REV 1 LIMITE.

Council of the European Union. (2021c, January 25). *Council Conclusions on Climate and Energy Diplomacy—Delivering on the External Dimension of the European Green Deal*, 3784th Meeting.

Dikaios, G. (2020, September 7–11). *International Implementation of EU Climate Norms: The Cases of the ICAO and the IMO*. Paper presented at the UACES 50th Annual Conference, Virtual Event.

Dikaios, G., & Blavoukos, S. (2023). Influencing the International Transport Regime Complex: The EU's Climate Action in ICAO and IMO. *Politics and Governance, 11*(2), 62–71.

Doussis, E. (2014). *Η Περιβαλλοντική Διακυβέρνηση σε Κρίση – Ρίο+20: Υποσχέσεις με Αβέβαιη Εφαρμογή* [Environmental Governance in Crisis – Rio+20: Promises with Uncertain Implementation]. Papazisi Publishers.

Earsom, J., & Delreux, T. (2021). A Nice Tailwind: The EU's Goal Achievement at the IMO Initial Strategy. *Politics and Governance*, *9*(3), 401–411.

Erling, U. M. (2017). International Aviation Emissions Under International Civil Aviation Organization's Global Market Based Measure: Ready for Offsetting? *Air & Space Law*, *42*(1), 1–12.

European Commission. (2013). *Communication from the Commission to the European Parliament, the Council, the European Economic and Social Committee and the Committee of the Regions—Integrating Maritime Transport Emissions in the EU's Greenhouse Gas Reduction Policies*, COM(2013) 479 final.

European Commission. (2016). *Communication from the Commission to the European Parliament, the Council, the European Economic and Social Committee and the Committee of the Regions—A European Strategy for Low-Emission Mobility*, COM(2016) 501 final.

European Commission. (2019). *Communication from the Commission to the European Parliament, the European Council, the Council, the European Economic and Social Committee and the Committee of the Regions: The European Green Deal*, COM(2019) 640 final.

European Commission. (2020). *Communication from the Commission to the European Parliament, the Council, the European Economic and Social Committee and the Committee of the Regions: Sustainable and Smart Mobility Strategy—Putting European Transport on Track for future*, COM(2020) 789 final.

European Commission. (2021). *Communication from the Commission to the European Parliament, the Council, the European Economic and Social Committee and the Committee of the Regions—'Fit for 55': Delivering the EU's 2030 Climate Target on the Way to Climate Neutrality*, COM(2021) 550 final.

European Commission. (2022a). *Reducing Emissions from Aviation*. https://ec.europa.eu/clima/eu-action/transport-emissions/reducing-emissions-aviation_en. Last accessed: 28 May 2022.

European Commission. (2022b). *Reducing Emissions from the Shipping Sector*. https://ec.europa.eu/clima/eu-action/transport-emissions/reducing-emissions-shipping-sector_en. Last accessed: 28 May 2022.

European Parliament. (2018a, March). *IMO's Challenges on the Route to Decarbonising International Shipping: Key Issues at Stake at the 72nd Session of the IMO Marine Environment Protection Committee (MEPC 72)*. Briefing.

European Parliament. (2018b). *Resolution of 3 July 2018 on Climate Diplomacy*. 2017/2272(INI).

European Parliament. (2020, September). *Greenhouse Gas Emissions from Shipping: Waiting for Concrete Progress at IMO Level*. Briefing.

European Parliament. (2021). *2030 Climate Target Plan: Extension of European Emission Trading System (ETS) to Transport Emissions*. In-Depth Analysis, ENVI Workshop Proceedings.

European Union Aviation Safety Agency. (2019). *European Aviation Environmental Report 2019*. https://doi.org/10.2822/309946

Gehring, M. W., & Robb, C. (2018). Sustainable Development and Emission Trading: The EU Perspective. In A. L. C. De Mestral, P. P. Fitzgerald, & T. Ahmad (Eds.), *Sustainable Development, International Aviation, and Treaty Implementation* (pp. 83–107). Cambridge University Press.

Harrison, J. (2017). *Saving the Ocean Through Law: The International Legal Framework for the Protection of the Marine Environment*. Oxford University Press.

ICAO. (2013). *38th Session of the ICAO Assembly, 24 September to 4 October 2013, Montréal: List of Delegates No. 10.0.*

ICAO. (2016). *39th Session of the ICAO Assembly, 27 September to 7 October 2016, Montréal: List of Delegates as of 06-Oct-2016 16:33 hrs.*

ICAO. (2019a). *40th Session of the ICAO Assembly, 24 September to 4 October 2019, Montréal: List of Delegates as of 11-Oct-2019 16:16 hrs.*

ICAO. (2019b). *Project Final Report: Capacity Building for CO_2 Mitigation from International Aviation*. ICAO.

ICAO. (2022, March). *Report on the Feasibility of a Long-term Aspirational Goal (LTAG) for International Civil Aviation CO_2 Emission Reductions*. https://www.icao.int/environmental-protection/LTAG/Documents/REPORT%20ON%20THE%20FEASIBILITY%20OF%20A%20LONG-TERM%20ASPIRATIONAL%20GOAL_en.pdf. Last accessed: 12 April 2022.

IMO. (2018a). *Data Collection System for Fuel Oil Consumption of Ships*. http://www.imo.org/en/OurWork/Environment/PollutionPrevention/AirPollution/Pages/Data-Collection-System.aspx. Last accessed: 13 May 2019.

IMO. (2018b, April 16). *List of Participants*, MEPC 72/INF.1.

IMO (2018c, October 29). *List of Participants*, MEPC 73/INF.1.

IMO. (2019, May 21). *List of Participants*, MEPC 74/INF.1.

IMO. (2021, July 6). *List of Participants*, MEPC 76/INF.1.

Jørgensen, K. E., & Laatikainen, K. V. (2004, 24–26 June). *The EU @ the UN: Multilateralism in a New Key?* Paper presented at the Second Pan-European Conference on EU Politics, Bologna.

Katsoulas, S. (2022). *The United States and Greek-Turkish Relations: The Guardian's Dilemma*. Routledge.

Kleistra, Y., & Van Willigen, N. (2015). Evaluating the Impact of EU Diplomacy: Pitfalls and Challenges. In J. A. Koops & G. Macaj (Eds.), *The European Union as a Diplomatic Actor* (pp. 52–68). Palgrave Macmillan.

Kleistra, Y., & Van Willigen, N. (2010). Evaluating Diplomacy: A Mission Impossible? *Evaluation, 16*(2), 119–135.

Kopela, S. (2017). Making Ships Cleaner: Reducing Air Pollution from International Shipping. *Review of European, Comparative & International Environmental Law, 26*(3), 231–242.

Kulovesi, K. (2012). Climate Change in EU External Relations: Please Follow My Example (Or I Might Force You To). In E. Morgera (Ed.), *The External Environmental Policy of the European Union: EU and International Law Perspectives* (pp. 115–148). Cambridge University Press.

Lin, W. (2017). Aviation and Climate Change: Practising Green Governmentality across the North-South Devide. *Geopolitics, 22*(1), 129–150.

Lindenthal, A. (2014). Aviation and Climate Protection: EU Leadership within the International Civil Aviation Organization. *Environmental Politics, 23*(6), 1064–1081.

Manners, I. (2002). Normative Power Europe: A Contradiction in Terms? *Journal of Common Market Studies, 40*(2), 235–258.

Martinez Romera, B. (2018). *Regime Interaction and Climate Change: The Case of International Aviation and Maritime Transport*. Routledge.

Martinez Romera, B., & van Asselt, H. (2015). The International Regulation of Aviation Emissions: Putting Differential Treatment into Practice. *Journal of Environmental Law, 27*(2), 259–283.

Maurer, H., & Wright, N. (2020). A New Paradigm for EU Diplomacy? EU Council Negotiations in a Time of Physical Restrictions. *The Hague Journal of Diplomacy, 15*(4), 556–568.

Nengye, L., & Maes, F. (2012). Legal Constraints to the European Union's Accession to the International Maritime Organization. *Journal of Maritime Law & Commerce, 43*(2), 279–291.

Newburger, E. (2020, October 24). Joe Biden Calls Climate Change the 'Number One Issue Facing Humanity'. *CNBC*. https://www.cnbc.com/2020/10/24/joe-biden-climate-change-is-number-one-issue-facing-humanity.html. Last accessed: 31 May 2022.

Oberthür, S. (2006). The Climate Change Regime: Interactions with ICAO, IMO, and the EU Burden-Sharing Agreement. In S. Oberthür & T. Gehring (Eds.), *Institutional Interaction in Global Environmental Governance Synergy and Conflict among International and EU Policies* (pp. 53–77). MIT Press.

Official Journal of the European Union. (2011). *Council Decision of 31 March 2011 on the Signing, on Behalf of the Union, and Provisional Application of a Memorandum of Cooperation between the European Union and the International Civil Aviation Organization Providing a Framework for Enhanced Cooperation (2011/530/EU)*, L 232/1.

Official Journal of the European Union. (2018). *Directive (EU) 2018/410 of the European Parliament and of the Council of 14 March 2018 amending*

Directive 2003/87/EC to enhance cost-effective emission reductions and low-carbon investments, and Decision (EU) 2015/1814 (Text with EEA relevance), L 76/3.

Official Journal of the European Union. (2021). *Regulation (EU) 2021/1119 of the European Parliament and of the Council of 30 June 2021 establishing the framework for achieving climate neutrality and amending Regulations (EC) No 401/2009 and (EU) 2018/1999 ('European Climate Law')*, L 243/1.

Shi, Y. (2018). The Implications of the Paris Agreement for the Regulation of Greenhouse Gas Emissions from International Shipping. *Ocean Yearbook, 32*, 528–555.

Shi, Y., & Gullet, W. (2018). International Regulation on Low-Carbon Shipping for Climate Change Mitigation: Development, Challenges and Prospects. *Ocean Development & International Law, 49*(2), 134–156.

Sjursen, H. (2006a). The EU as a 'Normative' Power: How Can This Be? *Journal of European Public Policy, 13*(2), 235–251.

Sjursen, H. (2006b). What Kind of Power? *Journal of European Public Policy, 13*(2), 169–181.

Van Leeuwen, J. (2019). Capturing a Moving Target: Decarbonizing Shipping Through Informational Governance. In P. G. Harris (Ed.), *Climate Change and Ocean Governance: Politics and Policy for Threatened Seas* (pp. 376–392). Cambridge University Press.

Van Leeuwen, J. (2015). The Regionalization of Maritime Governance: Towards a Polycentric Governance System for Sustainable Shipping in the European Union. *Ocean & Coastal Management, 117*, 23–31.

Van Leeuwen, J., & Kern, K. (2013). The External Dimension of European Union Marine Governance: Institutional Interplay between the EU and the International Maritime Organization. *Global Environmental Politics, 13*(1), 69–87.

Van Willigen, N., & Kleistra, Y. (2013). Evaluating Diplomacy. In K. E. Jørgensen & K. V. Laatikainen (Eds.), *Routledge Handbook on the European Union and International Institutions: Performance, Policy, Power* (pp. 102–112). Routledge.

Veno, J. (2007). Flying the Unfriendly Skies: The European Union's New Proposal to Include Aviation in their Emissions Trading Scheme. *Journal of Air Law and Commerce, 27*(3), 659–687.

Whiteman, A. (2018, May 14). IMO Listens to Members of European Parliament to Act to Reduce Shipping Emissions. *Canadian Sailings*, p. 48.

CHAPTER 6

Concluding Remarks: Navigating in Uncharted Seas and Skies—Normative (without) Power

Does the European Union perform a normative climate diplomacy towards the International Maritime Organization and the International Civil Aviation Organization? Does the 'normal' of the IMO and the ICAO change because of the EU's normative climate diplomacy that has aimed them? In other words, does the EU have the normative power, through its diplomacy, to alter the business-as-usual function of the IMO and the ICAO regarding the mitigations of greenhouse gas emissions? Eventually, if the answer to the above questions is positive, does it mean that the EU is a normative power in the field of climate change?

This book showcases the strategically driven efforts of the EU as a whole to export its climate norm in these two IOs. It makes the argument that if it weren't for the EU, these two Organizations would not have been active in negotiating on climate change to such an extent. It also shows that the internal diplomacy, external diplomacy and diplomacy evaluation of the EU constitute a three-part setup, which has enabled the performance of the *EU normative climate diplomacy*. More specifically, the book establishes that particular mechanisms/factors should be present for a negotiating position to be drafted and embedded in the EU's and the EU Member States' diplomatic apparatus, which will—*itaque*—allow for the successful externalization of this position.

© The Author(s), under exclusive license to Springer Nature Switzerland AG 2024

G. Dikaios, *EU Climate Diplomacy Towards the IMO and ICAO*, Palgrave Studies in European Union Politics,
https://doi.org/10.1007/978-3-031-51123-3_6

After the Paris Agreement was concluded in 2015, the international sentiment in favor of measures for tackling climate change targets was boosted significantly. Even the IMO and the ICAO, two IOs that have been described as 'unfriendly institutions' in the climate efforts (Oberthür, 2006: 59), succumbed momentarily to the global trend towards adopting pro-climate strategies. On the one hand, the ICAO adopted the CORSIA in 2016, after a significantly heated discussions period during which the EU attempted to impose market-based measures to all international flights landing to and departing from EU airports with the aim to regulate the reduction of aircraft GHG emissions in a global scale. On the other, the IMO adopted the Initial Strategy in 2018 after a long fermentation process, during which the EU was deliberately signaling that it would enforce a regional scheme to control the mitigation of ships' GHG emissions if the Organization failed to act. Both these strategies were concluded, admittedly, in a large part because of the pressure the EU put on the two IOs to engage with one of its top priority issues.

The CORSIA and the Initial Strategy are both texts that do not outline the specifics for their implementation to be successful. This is a rather common occurrence in the overarching international climate regime under the UNFCCC, where every year a Conference of the Parties takes place in order to evaluate the past year and to take decisions for the way forward. The same principle applies to all IOs that hold plenaries periodically to examine the progress of their mandate and to agree on the next steps to be taken. In the case of climate change in the ICAO and the IMO, the yearly meetings of the responsible institutional bodies are of outmost importance for the developments that would allow for decisions to be taken to contribute to the target of the reduction of GHG emissions by international aviation and shipping, respectively. While both agreements were saluted with excitement by the EU, several issues arise when a closer look to their essence and conception is taken. As it happens, the CORSIA fails to foresee the reduction of GHG emissions by airplanes but, rather, transfers the burden to other sectors. The Initial Strategy is a declaratory agreement with an insufficient mitigation goal. António Guterres, the incumbent UN Secretary Generally, urged both IOs to step up their game and do more, as their current efforts follow the trodden path that increases their GHG emissions (herein, p. vi). Nevertheless, it would be wrong to completely disregard the two texts. Both the CORSIA and the Initial Strategy were the best possible agreements that could be achieved through diplomatic means in a highly contested and politicized framework. It is interesting to note that the two IOs have adopted opposite strategies, both 'half-made', having left out the component that it would make them sufficiently implementable. The ICAO has the scheme without the goal and the IMO has the goal without the scheme. With time, the international negotiations that take place in the two IOs, have reached a point where they now concentrate on how the deficiencies of their respective systems can be resolved.

Within this absurd and erratic context, the EU attempts to seize the opportunity to externalize its climate norm, exporting a climate 'normal' that it believes should be applied globally. However, as it can be expected, in order for the Union to perform as a normative power in international climate negotiations, an actual climate norm should be in place and the internal EU stakeholders should be in agreement and in order. Therefore, the internal constellation and diplomatic procedures which ensure that the EU climate norm is embedded or acquired by all—without exceptions—European negotiators is of utmost importance; it is a prerequisite for the external—normative—climate diplomacy to be performed.

1 Tipping Points in the EU's Internal Climate Diplomacy

As a multifaceted organization formed by a number of entities which can represent it internationally, the EU is often seen as both the parts and the sum of them. When it comes to the case at hand, the first notion that needs clarification as well as justification in order to conclude which EU actor(s) has the responsibility to form the negotiating positions and strategies towards the IMO and the ICAO in climate change issues, is that of *competence*. The reason for exploring competence stems from the labyrinthine institutional framework that is in operation within the EU, as well as the relations of the Union with the two IOs. The incremental engagement of the supranational level with issues otherwise governed by the national levels for many years (especially in the case of the IMO), in concordance with the observer-only status that the EU and the Commission have in these IOs (and which does not allow for formal representation and voting rights), create a complex scaffold regarding the policies to be promoted externally and who the promoter of these policies could be.

According to the theoretical approach that has been presented, as well as to the EU's legal framework and case law, the Commission has exclusive competence on climate change aspects for international transport policies. However, this admission remains under severe scrutiny from the EU Member States who, in a constant strive to maintain as much power as possible on the issue at hand, keep the power equilibrium in constant flux. In an endeavor not to succumb to the individual Member States wills, the Commission maintains the view (emanating from the convergence of the factual elements regarding the exclusive competence) that it is the one who should be 'in charge' and puts significant efforts into establishing its power through the process of internal *coordination*.

Recent studies on the EU–IMO relations claim that the internal coordination is 'loose' and that the leading role at the international level is performed by the Council Presidency (Earsom & Delreux, 2021a; Laursen, 2020). This research is, though, based mainly on an article published ten years ago (Gulbrandsen, 2013). The case today, as showcased, proves rather different. The claims of 'looseness' seem to be rejected as intense internal EU diplomacy initiated by the Commission is observed. This action is aimed to secure the Commission's overarching position in climate shipping policies both internally and externally. The same finds application on the attempts that the Commission makes to determine the negotiating strategy of the EU Member States at the international level. This aggregated coordination might be the result of the inadequacy of the IMO to adopt the necessary measures foreseen in the Initial Strategy, which activates the EU's gut response to proceed with an aggressive diplomatic approach to make sure that the rules-based international order will deliver its promises. Therefore, by grabbing onto ambitious EU Member States which seem keen to accept a supranational solution to this subject, the Commission unfolds a penetrating coordination procedure to control all the Member States' positions (whether they are in favor or not of an EU common position in greening maritime affairs). The goal of this tactic is to, first, establish the Commission's jurisdiction and competence on this policy, second, to threaten the IMO with unilateral regional measures—although the IMO does not seem to bend over this pressure during the last couple of years—and third, to keep alive the *moral advantage* of being the *good force* that strives for a better (and sustainable) world.

In the aviation policy, the situation seems to be more balanced. A working arrangement between the Commission and the EU Member

States seems to be in place and operating quite well, although the former remains vigilant and is not being complacent. Instead, it constantly pushes hard to gain more power over the implementation of the international framework, namely the requirements of the CORSIA to be implemented through the EU ETS.[1]

The sluggish developments within the ICAO after the adoption of the CORSIA (Abeyratne, 2020), have led the Commission to obtain (again) a more stringent approach on the measures which it proposes for international aviation, sounding a warning notice both internally and externally for the next steps. Through this approach, once more, the Commission attempts to strengthen its position within the EU, as well as on the international level; at the same time, it enhances its self-description as one of (or even the) most green players globally.

The ongoing developments in the EU, through the adoption of legally binding EU Directives and Regulations focusing on tools that will reduce GHG emissions from both sectors (after the adoption of the European Climate Law), are expected to bring significant changes in the institutional EU framework. Due to this fact, concerns arise, on behalf of the EU Member States, that the Commission will gain more control over regulating GHG emissions internally, something that would have immediate consequences in the externalization of the EU voice. The cases under examination affirm the upheaval the Treaty of Lisbon brought into reality when the Member States agreed on the supranational jurisdiction of the EU's external relations, but subsequently use all available means trying to find ways to restrain the powers they delegated (cf. Van der Mei, 2016). Accordingly, and reversely, the Commission fights for the rights it has been given by the EU Treaties through extensive and intense policy- and decision-making, as well as through internal coordination, maintaining its position to be the institution in charge. Pending the finalization of the entirety of the new EU laws regarding climate change in European (and international) transport policies, the Commission strives for establishing its jurisdiction as much as possible so as for the latter to be reflected in the legal documents in the clearest way possible and to resolve the existing tensions (once and for all?) that emerge in a contested situation and threaten to jeopardize the European coherence and credibility globally.

[1] See e.g., Erling (2020) on how EU ETS and CORSIA can be reconciled.

It becomes evident that the first umbrella work assumption, i.e., *that the EU climate diplomacy is grounded on internal political procedures* is affirmed. This is the case as the internal EU climate diplomacy results in a coherent action by the EU as a whole and the analytical framework applied herein can be used to explain the ways the Commission attempts to appropriate more and more responsibilities, in order to enhance its positions in a certain (sub)policy field, confirming that the struggle of power between the national and the supranational levels is constant.

2 Caveats of the EU's External Climate Diplomacy and Its Unintended Consequences

The EU as a whole makes considerable efforts to convince third international organizations to adopt more climate-friendly policies. As the IOs, including the EU, are most often objectively seen as the bodies with regulatory capability (Abbott et al., 2015; Bradford, 2019; Hadjiyianni, 2019), the diplomacy that the EU performs towards other IOs is enshrined with substantial normative aspects, aiming at altering their business-as-usual plan and their 'normality'. For this reason, the EU employs an array of mechanisms that have been crystalized through the Normative Power Europe approach. Their realization is what diffuses its norm.

In both the IMO and the ICAO cases, the EU fulfills the expectations posed by the analytical framework in diffusing climate norms and, therefore, the second umbrella work assumption that the *EU climate diplomacy diffuses EU climate norms* is confirmed as well. The mechanisms proposed by the NPE approach sufficiently explain European diplomatic action and offer a useful analytical tool of an all-encompassing understanding of that action, while concurrently examining it in many and different levels. Thus, it can be argued that the EU climate diplomacy is normative in its core and, at the same time, that it disregards the classical Westphalian arrangement and allows for in-depth conceptualizations of how a non-state actor operates within the international diplomatic arena, giving prominence to both its internal and external action, which are deemed inseparable (again opposing the Westphalian status quo).

A challenge, though, emerges in regard to the level of success of the externalization of the EU climate norm of mitigating GHG emissions, and whether the aforementioned diffusion is indeed changing the 'normality' of the receiver IOs. Having shown that the diffusion of norms has not managed to produce any results in favor of the EU's interests yet or, more generally, in accordance with the (EU's) climate norm of mitigating GHG emissions, our initial approach of detaching the mechanisms of the NPE framework from the final outcome seems to be a valid argument. The reason is that although the EU performs a normative climate diplomacy and satisfies all the mechanisms put forward by the theory that would make one consider that the EU is indeed a normative power, it does not manage to successfully alter the 'norm-box' of the IOs. This said, *normative without power* refers to the conceptual basis of the NPE theory, challenging—in essence—its ability to test whether an actor has or is a normative power by examining its practice/activity. Moreover, the research results highlight the importance of the level of influence an actor needs to possess in order to be a normative power. Without the presence of the EU, developments towards reducing GHG emissions by ships and aircrafts would not have even started; nevertheless, the systems in place seem to underperform and to contain empty words about their goals. Consequently, regardless of the constant and rigorous efforts of the EU to move the agenda forward and to pose substantial rules to achieve the aspired outcome, since the sentiment of the international community is against any changes, there seems to be no success in fulfilling the EU's endeavors to be a normative power and to alter the 'normal' of the IOs. Taking into account that this unfavorable result followed the admitted 'wins' of concluding the Initial IMO Strategy and the CORSIA, the narrative of the EU being a normative power is emasculated further.

For the EU to be a normative *power* it should be able to change the normative core of the norm receiver, albeit the externalities. Consequently, the EU seems to be more resemblant of a *normative initiator*, a *normative leader*, or a *normative actor* but not a normative power. This observation also coincides with recent research on the European diplomacy added value (Maurer, 2022), which entangles diplomacy with actorness, arguing that the former consists of the means to exert the strategic objectives through adequate instruments which emanate from the latter. Thus, through performing normative diplomacy, the EU can be conceptualized as indeed a normative actor.

Despite the deficiencies that the external normative climate diplomacy displays, as has been shown in the case studies, a new approach on Normative Power emerges. It becomes evident that the EU does not only attempt to influence the exterior with its norm(s), but also its interior. The significant influence of an IO, even of a supranational hybrid as the

EU, in the decisions and behavior of its Member States comes as a sharp innovation in Westphalic relation, since it challenges—at the same time—state sovereignty. This last observation strongly empowers the normative supranational identity of the IO under consideration.

Theorizing, it becomes evident that the six mechanisms of externalizing a norm, as devised by Manners, can also be applied to the EU's internal level. The reasons are two: first, the EU Member States are concurrently IMO and ICAO Member States and, therefore, they get influenced by norms promoted by the EU institutions at the international level. Second, the EU Member States also get influenced internally by norms constructed in the European (supranational) level. Indeed, the NPE mechanisms seem to be also applicable to the internalization of the European norms at the level of the Member States, complementary and along—of course—with the internal practices analyzed herein.

In a nutshell, contagion and informational diffusion function internally in the same way as they do at the international level; procedural diffusion reflects the high institutionalization that exists between the EU institutions and the EU Members States; overt diffusion mirrors the internal coordination processes, while (internal) transference presents similar concerns with the claims that the third parties have for investments, i.e., EU Member States seem to be more receptive and open to support what the supranational level wants when the EU alleviates their financial burden and offers them incentives for the relevant action.[2]

When compared to the international level, the notable difference is that the internalization of the (supranational) EU climate norm in the international transport policy seems to have been successfully engulfed by the receiving actors, namely the EU Member States. The underlying logic of the cultural filter—which is explained through learning and adaptation—has led the Member States to alter their 'normal' to either incorporate climate concerns and strategies in their internal and external policies or to adjust their climate policies to the ones promoted by the supranational level. Moreover, this feature explains the external coherence that the EU as a whole performs, as well as the power the EU's institutions have over the positions of the EU Member States, leading to the argument that the EU is indeed a normative power, but in a completely different way than what was anticipated: it is a normative power for its internal area. As

[2] As also noted by some of the interviewees (Interviews no. 24, 33, 37).

such, the EU can be perceived as a single actor in the international system, while—at the same time—highlighting its state-functions. In turn, it also gives prominence to the discussion around the EU being a (supranational) state. And, although the effort of the EU to impose on the IMO and the ICAO to act according to its climate norm(s) has yet to bear fruits, the internal coherence that has been achieved on the way creates the promise that this goal will be eventually achieved.

3 The Past, the Present...

Commissioner Violeta Bulc signposts four reasons that allowed for successful outcomes during her mandate in both the IMO and the ICAO. These reasons go beyond what has been presented so far in the book. The first revolves around the fact that the overall awareness regarding the issues at stake was finally high enough to allow for the development of a comprehensive policy and a coherent position that, due to the lack of sufficient available data, previous Commissions did not have the chance to proceed with. The second, which is mentioned as a very personal observation of hers, is that the techniques that were applied by the predecessor schemes resembled too much to imperialistic behavior in general, pushing a dualist idea in the sense of the rationale *either you are with us or you are against us*, towards the IOs which created conflicts, dismay and unwillingness for establishing cooperations. Third, participatory models and inclusive approaches were introduced and started being implemented in the pursuit of collective action and of agreeing to a common denominator. Fourth, the Commission sent a very clear message, both internally and externally, that it wanted these agreements to be signed, because both sectors needed to take big steps in order to contribute to a global challenge.

Of course, one can oppose or—at least—scrutinize what Bulc claims, but the fact that the IMO Initial Strategy and the CORSIA were adopted with the EU playing an instrumental role to that end cannot be ignored. In spite of that observation, the course of action, since then, has shown a relative distancing to and from what the Commissioner noted as overarching reasons of success. Apart from the very clear message for more robust action towards the greening of the world that the current Commission signals, it seems that the latter estranges itself from more cooperative action both internally and externally.

Internally, an empowerment of the supranational level is in progress, which will lead to the diminishing of power of the EU Member States and to the delegation of additional powers to the European institutions. Although the interviewees coming from the EU Member States (in their overwhelming majority Ministry officers), pray not for such a development, as their role would be restricted to the management of the European policy and leave them with limited space for action, it seems that the upper echelons have made their decision concerning the political agreement. Subsequently, a gradual fade out of the current status quo of national officers who are able to produce policy outcomes in the field of transport policies regarding climate change policy directions will occur, validating both the competence and the coordination creep of the Commission in its (successful after all) attempts to establish its jurisdiction in an additional policy field. Consequentially, the internal competence of the supranational level leads to the looting of the external competence. This reaffirms the long-standing stratagem of functioning of the IOs, which allows them to extend their areas of responsibilities in fields which will facilitate the exercise of their explicitly delegated sectors, through the notion of implied powers.

The internal front's enhancement, however, does not lead to the respective enhancement of the external. This could be attributed to the unfriendly international environment and the widespread reluctancy of engaging with climate change measures in general by posing restrictions to the business-as-usual transport modes, in particular. Another factor could also be the observance of several of the interviewees that there is a slight deterioration of 'knowledge' and 'experience' on the matter possessed by the current Commission, which leads the latter to weaker and more superficial positions (Interviews no. 39, 40, 48). This might be the result of the reinstatement of more aggressive approaches that build mostly on 'big words' and move towards unilateral—internal—actions that will have external—intended—consequences, not to be attained via international negotiations and cooperative efforts, signaling the return of the *us versus them* mentality. For the EU Member States officials this is a rather undesirable development as they understand the need for being flexible and open to other points of view in order to achieve an agreement. They understand how through the enhancement of the internal supranational power they will have limited flexibility during the international negotiations and their individual role will be downgraded to the externalization of the common European position, without space for deviations or for inclusion of personal/national nuances during the international policymaking processes. Once again, this contributes to the argument that the normative power of the climate EU diplomacy is first and foremost impacting the behavior and the activity of the EU Member States, so that their consequent external action will influence the international level in regimes where the supranational level cannot be represented properly.

Nevertheless, both the present situation and the findings of the book at hand contradict the essence of the Normative Power Approach in being 'a holistic approach to [...] changing the planetary politics in the 21st century' (Manners, 2021: 69). Instead, they do justify it as a complementary theoretical explanation of the ways a (rather coercive) mode of Europeanization takes place. They affirm that the Normative Power Approach is a useful analytical tool and 'a holistic approach to studying' the EU's external action and diplomacy in the twenty-first century's contested global governance in a comprehensive manner.

4 ...and the Future

The promotion of strategic unilateral actions aiming to hold third actors accountable for not contributing their part to the global need of combating climate change points towards the acknowledgment of the failure to a large extent—for the time being—of the (external) normative power of the EU through its climate diplomacy. It remains to be seen whether coercive approaches will end up again being with ***kid gloves***[3] or—*au contraire*—will fulfil the expectations.

The turn towards a more strategic approach in the EU's climate diplomacy has been highlighted in the immediate aftermath of the conclusion of the Paris Agreement, when the introduction of strong geopolitical characteristics in the tackling of climate change (Oberthür, 2016), culminated to the development of the European Green Deal which puts emphasis on a holistic approach in climate policy that will intersect and integrate aspects of all internal and external EU policies. While the aim of the European Green Deal is to be the first step towards a grand climate strategy, it seems that this step has not yet been taken (Oberthür & Dupont, 2021).

[3] A term used by Birchfield (2015) for the freezing of the implementation of the EU ETS in international aviation.

The shift towards a more strategic roadmap in the EU's external relations and diplomacy has been in the works before the publication of the EU's Strategic Compass (European Union, 2022). This could also be attributed in part to the criticisms the Union has received for being *astrategic* (Cottey, 2020). The Compass was rushed to be presented in early 2022, due to the ongoing war in the European continent, a development which contests the post-Cold War status quo and introduces a period of great turbulence, instability and insecurity in the area. As the energy security of the EU is now a top priority, severe consequences on the course of the EU climate change policies and diplomacy will unfold, possibly challenging the current primary goals and targets. It is, thus, unavoidable for both theory and practice to delve into the inclusion of the strategic ramifications that all the external EU policies will experience in the years to come.

The effects of moving towards an overall strategic course of action encompass additional features to the action of the EU, some of them not necessarily apparent at first glance. This book focused in one aspect of a wide negotiating field, namely the internal and the external EU climate diplomacy. For strategic change to happen, with all key stakeholders being on board for transformations needed to contribute to the tackling of climate change, a holistic course of action needs to be agreed on and implemented. As the issue of climate change has become overly politicized and significant diplomatic efforts are utilized to balance the diverging and sometimes very strong (economic) interests, it is also interesting to take a closer look on what takes place within the *IOs that hate politics* (Louis & Maertens, 2021) and the way the shift to more politicized matters transpires within their frameworks. Besides, as it has methodically surfaced throughout the research at hand, no action on climate change would have taken place without the (peculiar) EU presence in the IMO and the ICAO.

During her interview, Violeta Bulc made a statement that sums up the situation. She eloquently argued that 'politics is the last pilar of society that has not reinvented itself'. For achieving climate targets, it becomes obvious that the hope for a better future lies with the evolvement of politics. The European Union needs to continue to be the *force for good* it has aspired to be and be the leader it can be. One actor does not a change make.

μία χελιδὼν ἔαρ οὐ ποιεῖ
Αριστοτέλους Ηθικά Νικομάχεια.
one swallow does not a summer make
Aristotle's Nicomachean Ethics.

List of Interviews

Interview no. 24, Advocate of the industry of an EU Member State, phone call, 17 October 2021.
Interview no. 33, EU Member State official, virtual call, 24 November 2021.
Interview no. 37, EU Member State official, virtual call, 1 December 2021.
Interview no. 39, EU Member State official, virtual call, 3 December 2021.
Interview no. 40, European Commission official, virtual call, 3 December 2021.
Interview no. 48, Violeta Bulc, European Commissioner of Transport (2014–19), virtual call, 20 December 2021.

References

Abbott, K. W., Genschel, P., Snidal, D., & Zangl, B. (Eds.). (2015). *International Organizations as Orchestrators*. Cambridge University Press.

Abeyratne, R. (2020). The Outcome of the 40th ICAO Assembly: A New Look at ICAO? *Air & Space Law, 45*(1), 81–96.

Birchfield, V. (2015). Coercion with Kid Gloves? The European Union's Role in Shaping a Global Regulatory Framework for Aviation Emissions. *Journal of European Public Policy, 22*(9), 1276–1294.

Bradford, A. (2019). *The Brussels Effect: How the European Union Rules the World*. Oxford University Press.

Cottey, A. (2020). Astrategic Europe. *Journal of Common Market Studies, 58*(2), 276–291.

Earsom, J., & Delreux, T. (2021a). A Nice Tailwind: The EU's Goal Achievement at the IMO Initial Strategy. *Politics and Governance, 9*(3), 401–411.

Erling, U. M. (2020). How to Reconcile the European Union Emissions Trading System (EU ETS) for Aviation with the Carbon Offsetting and Reduction Scheme for International Aviation (CORSIA)? *Air & Space Law, 43*(4&5), 371–386.

European Union. (2022, March). *A Strategic Compass for Security and Defence: For A European Union that Protects Its Citizens, Values and Interests and Contributes to International Peace and Security*. https://www.eeas.europa.eu/sites/default/files/documents/strategic_compass_en3_web.pdf. Last accessed: 12 April 2022.

Gulbrandsen, C. (2013). Navigating from Conflict to Working Arrangement: EU Coordination in the International Maritime Organization. *Journal of European Integration, 35*(7), 749–765.

Hadjiyianni, I. (2019). *The EU as A Global Regulator for Environmental Protection: A Legitimacy Perspective*. Hart Publishing.

Laursen, F. (2020). *The Development of the EU as A Sea-Policy Actor: Fish, Ships and Navies*. Edward Elgar Publishing.

Louis, M., & Maertens, L. (2021). *Why International Organizations Hate Politics: Depoliticizing the World*. Routledge.

Manners, I. (2021). Normative Power Approach to European Union External Action. In S. Gstöhl & S. Schunz (Eds.), *The External Action of the European Union: Concepts, Approaches, Theories* (pp. 69–84). Red Globe Press.

Maurer, H. (2022). The Added Value of European Diplomacy for EU Regional and International Actorness. In M. R. Freire, P. D. Lopes, D. Nascimento, & L. Simão (Eds.), *EU Global Actorness in a World of Contested Leadership: Policies, Instruments and Perceptions* (pp. 35–52). Palgrave Macmillan.

Oberthür, S. (2006). The Climate Change Regime: Interactions with ICAO, IMO, and the EU Burden-Sharing Agreement. In S. Oberthür & T. Gehring (Eds.), *Institutional Interaction in Global Environmental Governance Synergy and Conflict among International and EU Policies* (pp. 53–77). MIT Press.

Oberthür, S. (2016). Where to Go from Paris? The European Union in Climate Geopolitics. *Global Affairs, 2*(2), 119–130.

Oberthür, S., & Dupont, C. (2021). The European Union's International Climate Leadership: Towards A Grand Climate Strategy? *Journal of European Public Policy, 28*(7), 1095–1114.

Van der Mei, A. P. (2016). Case Note: EU External Relations and Internal Inter-Institutional Conflicts: The Battlefield of Article 218 TFEU. *Maastricht Journal of European and Comparative Law, 23*(6), 1051–1076.

List of Interviews

Interview no. 1, Advocate of the industry of an EU Member State, phone call, 31 March 2019.
Interview no. 2, Advocate of the industry of an EU Member State, phone call, 12 April 2019.
Interview no. 3, EU Member State official, phone call, 15 April 2019.
Interview no. 4, EU Member State official, phone call, 16 April 2019.
Interview no. 5, EU Member State official, phone call, 16 April 2019.
Interview no. 6, EU Member State official, phone call, 25 April 2019.
Interview no. 7, EU Member State official, phone call, 25 April 2019.
Interview no. 8, EU Member State official, phone call, 25 April 2019.
Interview no. 9, EU Member State official, phone call, 30 April 2019.
Interview no. 10, EU Member State official, phone call, 30 April 2019.
Interview no. 11, EU Member State official, email exchanges, 30 April 2019.
Interview no. 12, EU Member State official, phone call, 2 May 2019.
Interview no. 13, EU Member State official, phone call, 3 May 2019.
Interview no. 14, Advocate of the industry of an EU Member State, phone call, 7 May 2019.
Interview no. 15, European Parliament official, phone call, 7 May 2019.
Interview no. 16, EU Member State official, email exchanges, 10 May 2019.
Interview no. 17, EU Member State official, phone call, 15 May 2019.
Interview no. 18, European Commission official, phone call, 22 May 2019.
Interview no. 19, European Commission official, phone call, 22 May 2019.
Interview no. 20, EU Member State official, phone call, 22 May 2019.
Interview no. 21, EU Member State official, phone call, 22 May 2019.

© The Editor(s) (if applicable) and The Author(s), under exclusive license to Springer Nature Switzerland AG 2024

G. Dikaios, *EU Climate Diplomacy Towards the IMO and ICAO*, Palgrave Studies in European Union Politics, https://doi.org/10.1007/978-3-031-51123-3

Interview no. 22, EU Member State official, phone call, 24 May 2019.
Interview no. 23, EU Member State official, phone call, 27 May 2019.
Interview no. 24, Advocate of the industry of an EU Member State, phone call, 17 October 2021.
Interview no. 25, EU Member State official, virtual call, 2 November 2021.
Interview no. 26, Advocate of the industry of an EU Member State, virtual call, 10 November 2021.
Interview no. 27, EU Member State official, virtual call, 11 November 2021.
Interview no. 28, EU Member State official, virtual call, 16 November 2021.
Interview no. 29, EU Member State official, virtual call, 17 November 2021.
Interview no. 30, EU Member State official, virtual call, 17 November 2021.
Interview no. 31, EU Member State official, virtual call, 17 November 2021.
Interview no. 32, EU Member State official, virtual call, 18 November 2021.
Interview no. 33, EU Member State official, virtual call, 24 November 2021.
Interview no. 34, EU Member State official, virtual call, 26 November 2021.
Interview no. 35, European Parliament official, virtual call, 26 November 2021.
Interview no. 36, EU Member State official, virtual call, 30 November 2021.
Interview no. 37, EU Member State official, virtual call, 1 December 2021.
Interview no. 38, EU Member State official, virtual call, 1 December 2021.
Interview no. 39, EU Member State official, virtual call, 3 December 2021.
Interview no. 40, European Commission official, virtual call, 3 December 2021.
Interview no. 41, EU Member State official, virtual call, 3 December 2021.
Interview no. 42, EU Member State official, virtual call, 3 December 2021.
Interview no. 43, EU Member State official, virtual call, 6 December 2021.
Interview no. 44, EU Member State official, virtual call, 7 December 2021.
Interview no. 45, EU Member State official, virtual call, 7 December 2021.
Interview no. 46, EU Member State official, virtual call, 8 December 2021.
Interview no. 47, European Commission official, virtual call, 8 December 2021.
Interview no. 48, Violeta Bulc, European Commissioner of Transport (2014–19), virtual call, 20 December 2021.
Interview no. 49, Advocate of the industry of an EU Member State, phone call, 3 January 2022.
Interview no. 50, EU Member State officials, virtual call, 4 January 2022.
Interview no. 51, EU Member State official, virtual call, 10 January 2022.
Interview no. 52, European Commission official, virtual call, 24 January 2022.
Interview no. 53, European Parliament official, in person, 31 January 2022.
Interview no. 54, European Commission official, virtual call, 10 February 2022.
Origin of EU Member States officials and advocates of the industry interviewed: Austria, Belgium, Cyprus, Denmark, Estonia, Finland, Germany, Greece, Italy, Latvia, Luxemburg, Malta, Netherlands, Poland, Spain, Sweden, United Kingdom.

References

Abbott, K. W., Genschel, P., Snidal, D., & Zangl, B. (Eds.). (2015). *International Organizations as Orchestrators*. Cambridge University Press.

Abeyratne, R. (2017). Outcome of the 39th Session of the International Civil Aviation Organization Assembly. *Air & Space Law, 42*(1), 13–28.

Abeyratne, R. (2020). The Outcome of the 40th ICAO Assembly: A New Look at ICAO? *Air & Space Law, 45*(1), 81–96.

Adelle, C., Biedenkopf, K., & Torney, D. (Eds.). (2018). *European Union External Environmental Policy: Rules, Regulation and Governance beyond Borders*. Palgrave Macmillan.

Adler-Nissen, R. (2014). Symbolic Power in European Diplomacy: The Struggle between National Foreign Services and the EU's External Action Service. *Review of International Studies, 40*(4), 657–681.

Aeschlimann, J., & Regan, M. (2017). *The GA Handbook: A Practical Guide to the United Nations General Assembly* (2nd ed.). Permanent Mission of Switzerland to the United Nations.

Afionis, S. (2017). *The European Union in International Climate Change Negotiations*. Routledge.

Afionis, S., & Stringer, L. (2012). European Union Leadership in Biofuels Regulation: Europe as a Normative Power? *Journal of Cleaner Production, 32*, 114–123.

Afionis, S., & Stringer, L. (2014). The Environment as a Strategic Priority in the European Union-Brazil Partnership: Is the EU Behaving as a Normative Power or Soft Imperialist? *International Environmental Agreements, 14*, 47–64.

© The Editor(s) (if applicable) and The Author(s), under exclusive license to Springer Nature Switzerland AG 2024

G. Dikaios, *EU Climate Diplomacy Towards the IMO and ICAO*, Palgrave Studies in European Union Politics, https://doi.org/10.1007/978-3-031-51123-3

Agarwal, S. K. (2009). Mitigating Greenhous Gas (GHG) Emissions from International Shipping in Post-Kyoto Climate Policy: Legal Issues and Challenges. *Maritime Affairs: Journal of the National Maritime Foundation of India, 5*(1), 73–96.

Amanatidis, G., & Friedl, R. (2004). Science Related to Atmospheric Effects of Aircraft Emissions Continue to Mature. *ICAO Journal, 59*(5), 14–15.

ATAG. (2020). *Aviation Benefits Beyond Borders: Powering Global Economic Growth, Employment, Trade Links, Tourism and Support for Sustainable Development through Air Transport, Despite Global Crisis*. Air Transport Action Group.

Avrami, L. (2015). *Η πολιτική της Ευρωπαϊκής Ένωσης για την κλιματική αλλαγή: η συμμόρφωση ως παράγοντας διαμόρφωσης αποτελεσματικών εθνικών πολιτικών* [EU Climate Change Policy: Compliance as a Prerequisite for Domestic Policy Effectiveness] [PhD Thesis]. National and Kapodistrian University of Athens. http://hdl.handle.net/10442/hedi/36213. Last accessed: 21 February 2021.

Bäckstrand, K., & Elgström, O. (2013). The EU's Role in Climate Change Negotiations: From Leader to 'Leadiator.' *Journal of European Public Policy, 20*(10), 1369–1386.

Baltag, D. (2018). EU External Representation Post-Lisbon: The Performance of EU Diplomacy in Belarus, Moldova and Ukraine. *The Hague Journal of Diplomacy, 13*(1), 75–96.

Baltag, D., & Smith, M. (2015). EU and Member State Diplomacies in Moldova and Ukraine: Examining EU Diplomatic Performance Post-Lisbon. *European Integration Online Papers*, Vol. 19, article 5.

Barros, X. (2012). EU Counterterrorism and Aviation Security: Supranational Rules but Intergovernmental Politics? *European Foreign Affairs Review, 17*(Special Issue), 53–69.

Behrens, M., & Janusch, H. (2012). 'Great Normative Power': The European and American Trade Approach with Chile and Mexico. *European Foreign Affairs Review, 17*(3), 367–386.

Belis, D., Joffe, P., Kerremans, B., & Qi, Y. (2015). China, the United States and the European Union: Multiple Bilateralism and Prospects for a New Climate Change Diplomacy. *Climate Change Law Review, 3*, 203–218.

Bengtsson, R. (2015). The EU and Global Economic Governance: Playing the Role of a Global Leader? In A. Bjorkdahl et al. (Eds.), *Importing EU Norms: Conceptual Framework and Empirical Findings* (pp. 39–54). Springer.

Berg, C. (2021). The External Representation of the European Union in the International Maritime Organization: A Question of Labelling rather than of Competence. *EU Law Analysis: Expert Insight into EU Law Developments*. http://eulawanalysis.blogspot.com/2021/12/. Last accessed: 30 March 2022.

Biedenkopf, K. (2017). Relations between International Organizations in Combating Climate Change. In R. Biermann & J. A. Koops (Eds.), *The Palgrave Handbook of Inter-organizational Relations in World Politics* (pp. 649–676). Palgrave Macmillan.

Biedenkopf, K., & Petri, F. (2019). EU Delegations in European Union Climate Diplomacy: The Role of Links to Brussels, Individuals and Country Contexts. *Journal of European Integration, 41*(1), 47–63.

Biedenkopf, K., & Petri, F. (2021). The European External Action Service and EU Climate Diplomacy: Coordinator and Supporter in Brussels and Beyond. *European Foreign Affairs Review, 26*(1), 71–86.

Birchfield, V. (2015). Coercion with Kid Gloves? The European Union's Role in Shaping a Global Regulatory Framework for Aviation Emissions. *Journal of European Public Policy, 22*(9), 1276–1294.

Birchfield, V. (2013). A Normative Power Europe Framework of Transnational Policy Formation. *Journal of European Public Policy, 20*(6), 907–922.

Blavoukos, S. (2015). Capturing the EU's International Performance: An Analytical Framework. In S. Blavoukos, D. Bourantonis, & C. Portela (Eds.), *The EU and the Non-Proliferation of Nuclear Weapons: Strategies, Policies, Actions* (pp. 12–30). Palgrave Macmillan.

Blavoukos, S., & Bourantonis, D. (2013, May 9–11). Struggling with Performance: An Analytical Framework for the EU International Interactions. *Paper prepared for the EUSA Thirteenth Biennial Conference.*

Blavoukos, S., & Bourantonis, D. (Eds.). (2017). *The EU in UN Politics: Actors, Processes and Performances*. Palgrave Macmillan.

Blavoukos, S., Bourantonis, D., & Tsakonas, P. (Eds.). (2019). *Εξωτερικές Σχέσεις της ΕΕ [The External Relations of the EU]*. I. Sideris Publications.

Blavoukos, S., Bourantonis, D., Galariotis, I., & Gianniou, M. (2016). The European Union's Visibility and Coherence at the United Nations General Assembly. *Global Affairs, 2*(1), 35–45.

Blokker, N. M. (2009). International Organization or Institutions, Implied Powers. In R. Wolfrum (Ed.), *Max Planck Encyclopedia of Public International Law*. Oxford University Press.

Bo, Y., Biedenkopf, K., & Chen, Z. (2016). Chinese and EU Climate and Energy Security Policy. In E. J. Kirchner, T. Christiansen, & H. Dorussen (Eds.), *Security Relations between China and the European Union: From Convergence to Cooperation?* (pp. 102–123). Cambridge University Press.

Bocse, A. M. (2020). The UK's Decision to Leave the European Union (Brexit) and Its Impact on the EU as a Climate Change Actor. *Climate Policy, 20*(2), 265–274.

Borrell, J. (2021, November 12). A Strategic Compass for Europe. *Project Syndicate*. https://www.project-syndicate.org/commentary/eu-strategic-compass-by-josep-borrell-2021-11. Last accessed: 22 March 2022.

Börzel, T. A., & Risse, T. (2012). When Europeanisation Meets Diffusion: Exploring New Territory. *West European Politics, 35*(1), 192–207.

Botta, M. (2014). Competition Policy: The EU and Global Networks. In G. Falkner & P. Müller (Eds.), *EU Policies in A Global Perspective: Shaping or Taking International Regimes?* (pp. 76–92). Routledge.

Bradford, A. (2019). *The Brussels Effect: How the European Union Rules the World*. Oxford University Press.

Branger, F., Lecuyer, O., & Quirion, P. (2015). The European Union Emissions Trading Scheme: Should We Throw the Flagship Out with the Bathwater? *Wires Climate Change, 6*(1), 9–16.

Brasseur, G., Amanatidis, G. T., & Angeletti, G. (Guest Eds.). (1998). European Scientific Assessment of the Atmospheric Effects of Aircraft Emissions. *Atmospheric Environoment, 32*(13), 2327–2422.

Braun, M. (2014). EU Climate Norms in East-Central Europe. *Journal of Common Market Studies, 52*(3), 445–460.

Brun, A. (2016). Conference Diplomacy: The Making of the Paris Agreement. *Politics and Governance, 4*(3), 115–123.

Buck, M. (2012). The EU's Representation in Multilateral Environmental Negotiations after Lisbon. In E. Morgera (Ed.), *The External Environmental Policy of the European Union: EU and International Law Perspectives* (pp. 76–95). Cambridge University Press.

Bulc, V. (2018, July 12). Shipping Is Delivering on Climate Change. *The Parliament Magazine*. https://www.theparliamentmagazine.eu/news/article/shipping-is-delivering-on-climate-change. Last accessed: 13 March 2022.

Bullock, S., Mason, J., & Larkin, A. (2022). The Urgent Case for Stronger Climate Targets for International Shipping. *Climate Policy, 22*(3), 301–309.

Callon, M. (1984). Some Elements of A Sociology of Translation: Domestication of the Scallops and the Fishermen of St Brieuc Bay. *The Sociological Review, 32*(1), 196–233.

Chaban, N., Knodt, M., & Verdun, A. (Guest Eds.). (2017). Special Issue: 'External images of the EU-Energy power Across the Globe'. *Comparative European Politics, 15*(1), 1–155.

Chamon, M. (2018). Implied Exclusive Powers in the ECJ's Post-Lisbon Jurisprudence: The Continued Development of the *ERTA* Doctrine. *Common Market Law Review, 55*(4), 1101–1142.

Chou, M. H., & Riddervold, M. (2015). The Unexpected Negotiator at the Table: How the European Commission's Expertise Informs Intergovernmental EU Policies. *Politics and Governance, 3*(1), 61–72.

Christensen, J. (2015a). Recruitment and Expertise in the European Commission. *West European Politics, 38*(3), 649–678.

Christensen, J. (2015b). The Organization of Professional Expertise in the European Commission. *Politics and Governance, 3*(1), 13–25.

Commission of the European Communities. (2001). *White Paper—European Transport Policy for 2010: Time to Decide*, COM(2001) 370 final.

Commission of the European Communities. (2007). *Communication from the Commission to the European Parliament, the Council, the European Economic and Social Committee and the Committee of the Regions: An Integrated Maritime Policy for the European Union*. COM(2007) 575 final.

Corbett, J., Ruwet, M., Xu, Y. C., & Weller, P. (2020). Climate Governance, Policy Entrepreneurs and Small States: Explaining Policy Change at the International Maritime Organization. *Environmental Politics, 29*(5), 825–844.

Corbett, J. J., & Winebrake, J. J. (2010). The Role of International Policy in Mitigating Global Shipping Emissions. *The Brown Journal of World Affairs, 16*(2), 143–158.

Costa, O. (2017). Assessing Influence between International Organizations. In R. Biermann & J. A. Koops (Eds.), *The Palgrave Handbook of Inter-organizational Relations in World Politics* (pp. 389–406). Palgrave Macmillan.

Cottey, A. (2020). Astrategic Europe. *Journal of Common Market Studies, 58*(2), 276–291.

Council of the European Union. (2009, October 21). *Council Conclusions on EU Position for the Copenhagen Climate Conference (7–18 December 2009)*, 2968th Environment Council Meeting.

Council of the European Union. (2011, July 18). *Council Conclusions on EU Climate Diplomacy*, 3106th Foreign Affairs Council Meeting.

Council of the European Union. (2013, June 24). *Council Conclusion on EU Climate Diplomacy*, Foreign Affairs Council Meeting.

Council of the European Union. (2016, February 15). *European Climate Diplomacy after COP21—Council Conclusions (15 February 2016)*, 3447th Meeting.

Council of the European Union. (2017, March 6). *Implementing the EU Global Strategy—Strengthening Synergies between EU Climate and Energy Diplomacies and Elements for Priorities for 2017—Council Conclusions (6 March 2017)*, 3525th meeting.

Council of the European Union. (2018, February 26). *Council Conclusions on Climate Diplomacy—Council Conclusions (26 February 2018)*, 3598th Meeting.

Council of the European Union. (2019a, February 18). *Climate Diplomacy—Council Conclusions (18 February 2019)*, 3673rd Meeting.

Council of the European Union. (2019b, November 11). *Sixth Meeting of the Intersessional Working Group on Reduction of GHG Emissions from Ships (ISWG-GHG 6) (London, 11–15 November 2019)—Non-paper from the Commission Drafted to Facilitate EU Co-ordination*. 13148/3/19 REV 3 LIMITE.

Council of the European Union. (2019c, September 10). *Union Position for the Reply to ICAO State Letter on CORSIA Sustainable Aviation Fuels (SL 19/55)—Endorsement*, 11886/19.

Council of the European Union. (2020a, June 19). *75th Session of the IMO Marine Environment Protection Committee (London, 30 March–3 April 2020)—EU Coordination*. 6419/2/20 REV 2 LIMITE.

Council of the European Union. (2020b, January 20). *Council Conclusion on Climate Diplomacy*, 3742nd Meeting.

Council of the European Union. (2020c, January 31). *IMO—Draft Union Submission to the Seventh Meeting of the Intersessional Working Group on Reduction of GHG Emissions from Ships of the IMO in London from 23 to 27 March 2020 on the Introduction of Life Cycle Guidelines to Estimate Well-To-Tank Greenhouse Gas (GHG) Emissions of Sustainable Alternative Fuels to Incentivise the Uptake of Sustainable Alternative Fuels at Global Level—Endorsement*, 5606/20.

Council of the European Union. (2020d, June 10). *Union Position on the Reply to ICAO State Letter ENV 6/6–20/53: Proposed Amendments to the ICAO Document 'CORSIA Sustainability Criteria for CORSIA Eligible Fuels', reference in Annex 16, Volume IV*, 8728/20.

Council of the European Union. (2021a, May 11). *76th Session of the IMO Marine Environment Protection Committee (10–17 June 2021)—EU Coordination*. 8277/1/21 REV 1 LIMITE.

Council of the European Union. (2021b, November 15). *77th Session of the IMO Marine Environment Protection Committee (22–26 November 2021)—EU Coordination*. 13527/1/21 REV 1 LIMITE.

Council of the European Union. (2021c, January 25). *Council Conclusions on Climate and Energy Diplomacy—Delivering on the External Dimension of the European Green Deal*, 3784th Meeting.

Court of Justice of the European Union. (2009). *Judgement of the Court (Second Chamber) of 12 February 2009. Commission of the European Communities v Hellenic Republic. Failure of a Member State to fulfil obligations—Article 10 EC, 71 EC and 80(2) EC—Maritime safety—Monitoring of ships and port facilities—International agreements—Division of powers between the Community and the Member States. Case C-45/07*. https://curia.europa.eu/juris/liste.jsf?language=en&num=C-45/07. Last accessed: 30 November 2022.

Cross, M. K. D. (2018). Partners at Paris? Climate Negotiations and Transatlantic Relations. *Journal of European Integration, 40*(5), 571–586.

Da Conceição-Heldt, E., & Meunier, S. (2014a). Speaking with a Single Voice: Internal Cohesiveness and External Effectiveness of the EU in Global Governance. *Journal of European Public Policy, 21*(7), 961–979.

Da Conceição-Heldt, E., & Meunier, S. (Guest Eds.). (2014b). Speaking with a Single Voice: The EU as An Effective Actor in Global Governance. *Journal of European Public Policy, 21*(7), 961–1083.

Dąbrowski, M. (2014). Transport Policy: EU as A Taker, Shaper or Shaker of the Global Civil Aviation Regime? In G. Falkner & P. Müller (Eds.), *EU Policies in a Global Perspective: Shaping or Taking International Regimes?* (pp. 130–148). Routledge.

Dahl, A. (2000). Competence and Subsidiarity: Legal Basis and Political Realities. In J. Gupta & M. Grubb (Eds.), *Climate Change and European Leadership: A Sustainable Role for Europe?* (pp. 203–220). Kluwer Academic Publishers.

Damro, C. (2006). EU-UN Environmental Relations: Shared Competence and Effective Multilateralism. In K. V. Laatikainen & K. E. Smith (Eds.), *The European Union at the United Nations: Intersecting Multilateralisms* (pp. 175–192). Palgrave Macmillan.

Damro, C. (2012). Market Power Europe. *Journal of European Public Policy, 19*(5), 682–699.

Damro, C., Gstöhl, S., & Schunz, S. (2018). *The European Union's Evolving External Engagement: Towards New Sectoral Diplomacies?* Routledge.

De Cock, G. (2011). The European Union as a Bilateral 'Norm Leader' on Climate Change vis-à-vis China. *European Foreign Affairs Review, 16*(1), 89–105.

De Paula Domingos, P. (2012). Fighting Climate Change in the Air: Lessons from the EU Directive on Global Aviation. *Revista Brasileira de Política Internacional, 55*(special edition), 70–87.

De Zutter, E. (2010). Normative Power Spotting: An Ontological and Methodological Appraisal. *Journal of European Public Policy, 17*(8), 1106–1127.

Deane, F., Huggins, A., & Karim, S. M. (2019). Measuring, Monitoring, Reporting and Verification of Shipping Emissions: Evaluating Transparency and Answerability. *Review of European, Comparative & International Environmental Law, 28*(3), 258–267.

DeCanio, S. J., & Fremstad, A. (2013). Game Theory and Climate Diplomacy. *Ecological Economics, 85*, 177–187.

Degrand Guillaud, A. (2009a). Actors and Mechanisms of EU Coordination at the UN. *European Foreign Affairs Review, 14*(3), 405–430.

Degrand Guillaud, A. (2009b). Characteristics of and Recommendations for EU Coordination at the UN. *European Foreign Affairs Review, 14*(4), 607–622.

Dejong, S. M. (2013). Hot Air and Hot Heads: An Examination of the Legal Arguments Surrounding the Extension of the European Union's Emissions Trading Scheme to Aviation. *Asian Journal of International Law, 3*(1), 163–188.

Del Sarto, R. A. (2016). Normative Empire Europe: The European Union, Its Borderlands, and the 'Arab Spring.' *Journal of Common Market Studies, 54*(2), 215–232.

Delegation of the European Union to Burundi. (2019). *EU Commissioner for Transport Meets with Seoul Mayor*. https://eeas.europa.eu/delegations/burundi/63685/eu-commissioner-transport-meets-seoul-mayor_zh-tw. Last accessed: 28 March 2022.

Delreux, T. (2006). The European Union in International Environmental Negotiations: A Legal Perspective on the Internal Decision-Making Process. *International Environmental Agreements: Politics, Law and Economics, 6*(3), 231–248.

Delreux, T. (2009). Cooperation and Control in the European Union: The Case of the European Union as International Environmental Negotiator. *Cooperation and Conflict: Journal of the Nordic International Studies Association, 44*(2), 189–208.

Delreux, T. (2011). *The EU as International Environmental Negotiator*. Ashgate.

Delreux, T. (2014). EU Actorness, Cohesiveness and Effectiveness in Environmental Affairs. *Journal of European Public Policy, 21*(7), 1017–1032.

Delreux, T., & Burns, C. (2019). Parliamentarizing a Politicized Policy: Understanding the Involvement of the European Parliament in UN Climate Negotiations. *Politics and Governance, 7*(3), 339–349.

Delreux, T., & Ohler, F. (2019). Climate Policy in the European Union Politics. *Oxford Research Encyclopedia: Politics*. https://doi.org/10.1093/acrefore/9780190228637.013.1097

Delreux, T., & Pirlot, P. (2017). The EU Performance in the UN Forum on Forests: An Analysis of the EU at UNFF11. In S. Blavoukos & D. Bourantonis (Eds.), *The EU in UN Politics: Actors, Processes and Performances* (pp. 187–207). Palgrave Macmillan.

Di Robilant, A. (2006). Genealogies of Soft Law. *The American Journal of Comparative Law, 54*(3), 499–554.

Diez, T. (2004). Constructing the Self and Changing Others: Reconsidering NPE. *Millennium: Journal of International Studies, 33*(3), 613–636.

Diez, T. (2013). Normative Power as Hegemony. *Cooperation and Conflict, 48*(2), 194–210.

Dikaios, G. (2020, September 7–11). *International Implementation of EU Climate Norms: The Cases of the ICAO and the IMO*. Paper presented at the UACES 50th Annual Conference, Virtual Event.

Dikaios, G., & Blavoukos, S. (2023). Influencing the International Transport Regime Complex: The EU's Climate Action in ICAO and IMO. *Politics and Governance, 11*(2), 62–71.

Dikaios, G., & Tsagkroni, V. (2021). Failing to Build a Network as Policy Entrepreneurs: Greek Politicians Negotiating with the EU during the First Quarter of SYRIZA in Government. *Contemporary Politics, 27*(5), 611–630.

Dirix, J., Peeters, W., & Sterckx, S. (2015). Is the EU ETS a Just Climate Policy? *New Political Economy, 20*(5), 702–724.

Doelle, M., & Chircop, A. (2019). Decarbonizing International Shipping: An Appraisal of the IMO's Initial Strategy. *Review of European, Comparative & International Environmental Law, 28*(3), 268–277.

Doussis, E. (2001). *Η Κοινοτική Πολιτική Περιβάλλοντος και η Επίδρασή της στην Περίπτωση της Ελλάδας* [The Community's Environmental Policy and Its Impact on the Greek Case]. Papazisi Publications.

Doussis, E. (2009). Le principle d'intégration. In P. Billet, M. Durousseau, G. Martin, & I. Trinquelle (Eds.), *Droit de l'environnement et protection de la santé* (pp. 255–260). L'Harmattan.

Doussis, E. (2014). *Η Περιβαλλοντική Διακυβέρνηση σε Κρίση – Ρίο+20: Υποσχέσεις με Αβέβαιη Εφαρμογή* [Environmental Governance in Crisis – Rio+20: Promises with Uncertain Implementation]. Papazisi Publishers.

Doussis, E. (2017). *Κλιματική Αλλαγή* [Climate Change]. Papadopoulos Publications.

Doussis, E. (2019). Η Εξωτερική Δράση της ΕΕ για το Περιβάλλον και το Παράδειγμα της Κλιματικής Αλλαγής [The External Action of the EU on the Environment and the Example of Climate Change]. In S. Blavoukos, D. Bourantonis, & P. Tsakonas (Eds.), *Εξωτερικές Σχέσεις της ΕΕ* [The External Relational of the EU] (pp. 303–320). I. Sideris Publications.

Doussis, E. (2020). *Διεθνές Δίκαιο και Διπλωματία της Κλιματικής Αλλαγής* [International Law and Climate Change Diplomacy]. Nomiki Vivliothiki.

Doussis, E. (2021, 30 August–3 September). *The Road to Climate Neutrality and Just Transition in Southern Europe: The Case of Greece*. Paper presented at the ECPR General Conference, Virtual Event.

Dreger, J. (2014). *The European Commission's Energy and Climate Policy: A Climate for Expertise?* Palgrave Macmillan.

Dupont, C., & Moore, B. (2019). Brexit and the EU in Global Climate Governance. *Politics and Governance, 7*(3), 51–61.

Dupont, C., & Oberthür, S. (Eds.). (2015). *Decarbonization in the European Union: Internal Policies and External Strategies*. Palgrave Macmillan.

Dupont, C., Oberthür, S., & Biedenkopf, K. (2018). Climate Change: Adapting to Evolving Internal and External Dynamics. In C. Adelle, K. Biedenkopf, & D. Torney (Eds.), *European Union External Environmental Policy: Rules, Regulation and Governance Beyond Borders* (pp. 105–124). Palgrave Macmillan.

Earsom, J., & Delreux, T. (2021a). A Nice Tailwind: The EU's Goal Achievement at the IMO Initial Strategy. *Politics and Governance, 9*(3), 401–411.

Earsom, J., & Delreux, T. (2021b). Evaluating EU Responsiveness to the Evolution of the International Regime Complex on Climate Change. *International Environmental Agreements: Politics, Law and Economics, 21*(4), 711–728.

Eckert, S. (2021). The European Green Deal and the EU's Regulatory Power in Times of Crisis. *Journal of Common Market Studies, 59*(S1), 81–91.

Elgström, O. (2016). *Legitimacy, Credibility and Coherence—Perceptions of EU Roles in Global Climate Change Negotiations*. EUI Working Paper, RSCAS 2016/06.

Ellerman, D. (2009). The EU's Emissions Trading Scheme: A Prototype Global System? *MIT Joint Program on the Science and Policy of Global Change*, Report no. 170.

Erling, U. M. (2020). How to Reconcile the European Union Emissions Trading System (EU ETS) for Aviation with the Carbon Offsetting and Reduction Scheme for International Aviation (CORSIA)? *Air & Space Law, 43*(4&5), 371–386.

Erling, U. M. (2017). International Aviation Emissions Under International Civil Aviation Organization's Global Market Based Measure: Ready for Offsetting? *Air & Space Law, 42*(1), 1–12.

euobserver. (2021, February 22). EU Preparing Airline Eco-Labels for 2022. *Euobserver*. https://euobserver.com/tickers/151001. Last accessed: 4 February 2022.

EUR-Lex. (2020). *Glossary of Summaries: Open Method of Coordination*. https://eur-lex.europa.eu/summary/glossary/open_method_coordination.html. Last accessed: 4 November 2020.

EUROCONTROL. (2022). *About Us*. https://www.eurocontrol.int/about-us. Last accessed: 29 May 2022.

European Civil Aviation Conference. (2022). *European Aviation and Environment Working Group (EAEG)*. https://www.ecac-ceac.org/activities/environment/european-aviation-and-environment-working-group-eaeg. Last accessed: 4 February 2022.

European Commission. (2011). *White Paper: Roadmap to a Single European Transport Area—Towards a Competitive and Resource Efficient Transport System*, COM(2011) 144 final.

European Commission. (2012). *Stopping the Clock of ETS and Aviation Emissions Following Last Week's International Civil Aviation Organization (ICAO) Council*. MEMO/12/854.

European Commission. (2013). *Communication from the Commission to the European Parliament, the Council, the European Economic and Social*

Committee and the Committee of the Regions—Integrating Maritime Transport Emissions in the EU's Greenhouse Gas Reduction Policies, COM(2013) 479 final.

European Commission. (2016). *Communication from the Commission to the European Parliament, the Council, the European Economic and Social Committee and the Committee of the Regions—A European Strategy for Low-Emission Mobility*, COM(2016) 501 final.

European Commission. (2019a). *Communication from the Commission to the European Parliament, the European Council, the Council, the European Economic and Social Committee and the Committee of the Regions: The European Green Deal*, COM(2019) 640 final.

European Commission. (2019b). *Decision of the President of the European Commission of 1 December 2019 on the Creation of Commissioners' Groups*, P(2019) 4.

European Commission. (2020). *Communication from the Commission to the European Parliament, the Council, the European Economic and Social Committee and the Committee of the Regions: Sustainable and Smart Mobility Strategy—Putting European Transport on Track for future*, COM(2020) 789 final.

European Commission. (2021). *Communication from the Commission to the European Parliament, the Council, the European Economic and Social Committee and the Committee of the Regions—'Fit for 55': Delivering the EU's 2030 Climate Target on the Way to Climate Neutrality*, COM(2021) 550 final.

European Commission. (2022a). *EU Emissions Trading System (EU ETS)*. https://ec.europa.eu/clima/eu-action/eu-emissions-trading-system-eu-ets_en. Last accessed: 4 April 2022.

European Commission. (2022b). *International Action on Climate Change*. https://ec.europa.eu/clima/eu-action/international-action-climate-change_en. Last accessed: 28 May 2022.

European Commission. (2022c). *Reducing Emissions from Aviation*. https://ec.europa.eu/clima/eu-action/transport-emissions/reducing-emissions-aviation_en. Last accessed: 28 May 2022.

European Commission. (2022d). *Reducing Emissions from the Shipping Sector*. https://ec.europa.eu/clima/eu-action/transport-emissions/reducing-emissions-shipping-sector_en. Last accessed: 28 May 2022.

European Commission. (2022e). *Revision for Phase 4 (2021–2030)*. https://ec.europa.eu/clima/eu-action/eu-emissions-trading-system-eu-ets/revision-phase-4-2021-2030_en. Last accessed: 31 May 2022.

European Commission. (2022f). *2020 Climate & Energy Package*. https://ec.europa.eu/clima/eu-action/climate-strategies-targets/2020-climate-energy-package_en. Last accessed: 31 May 2022.

European Council (1990) *Presidency Conclusions, Dublin, 25 and 26 June 1990*, SN 60/1/90.

European Parliament. (2018a, March). *IMO's Challenges on the Route to Decarbonising International Shipping: Key Issues at Stake at the 72nd Session of the IMO Marine Environment Protection Committee (MEPC 72)*. Briefing.

European Parliament. (2018b). *Resolution of 3 July 2018 on Climate Diplomacy*. 2017/2272(INI).

European Parliament. (2020, September). *Greenhouse Gas Emissions from Shipping: Waiting for Concrete Progress at IMO Level*. Briefing.

European Parliament. (2021). *2030 Climate Target Plan: Extension of European Emission Trading System (ETS) to Transport Emissions*. In-Depth Analysis, ENVI Workshop Proceedings.

European Parliament. (2022, February 28). *ReFuelEU Aviation and Maritime: ENVI Exchange on Opinions*. https://www.europarl.europa.eu/committees/en/refueleu-aviation-and-maritime-envi-exch/product-details/20220214CAN64971. Last accessed: 1 April 2022.

European Union. (2016, June). *Shared Vision, Common Action: A Stronger Europe—A Global Strategy for the European Union's Foreign and Security Policy*. http://www.eeas.europa.eu/archives/docs/top_stories/pdf/eugs_review_web.pdf. Last accessed: 13 March 2022.

European Union. (2022, March). *A Strategic Compass for Security and Defence: For A European Union that Protects Its Citizens, Values and Interests and Contributes to International Peace and Security*. https://www.eeas.europa.eu/sites/default/files/documents/strategic_compass_en3_web.pdf. Last accessed: 12 April 2022.

European Union Aviation Safety Agency. (2019). *European Aviation Environmental Report 2019*. https://doi.org/10.2822/309946

Ewers-Peters, N. M. (2018). The EU as An Interorganizational Influencer? *European Foreign Affairs Review, 23*(3), 365–380.

Fahey, E. (2012). The EU Emissions Trading Scheme and the Court of Justice: The 'High Politics' of Indirectly Promoting Global Standards. *German Law Review, 13*(11), 1247–1268.

Falkner, R. (2007). The Political Economy of 'Normative Power' Europe: EU Environmental Leadership in International Biotechnology Regulation. *Journal of European Public Policy, 14*(4), 507–526.

Field, C. B., Barros, V. R., Dokken, D. J., Mach, K. J., Mastrandrea, M. D., et al. (Eds.). (2014). *Climate Change 2014: Impacts, Adaptation, and Vulnerability*. Cambridge University Press.

Finnemore, M., & Sikkink, K. (1998). International Norm Dynamics and Political Change. *International Organization, 52*(4), 887–917.

Fischer, S., & Geden, O. (2015). The Changing Role in International Negotiations in EU Climate Policy. *The International Spectator: Italian Journal of International Affairs, 50*(1), 1–7.

Forsberg, T. (2011). Normative Power Europe, Once Again: A Conceptual Analysis of an Ideal Type. *Journal of Common Market Studies, 49*(6), 1183–1204.

Frick, M., & Schulz, S. (2014). Climate Change Politics and Policies: Germany and the European Union in the Climate Negotiations. In W. Hofmeister & P. Rueppel (Eds.), *Climate Change Diplomacy: The Way Forward for Asia and Europe* (pp. 37–52). Kondrad Adenauer Stiftung; East Asian Institute; European Union Centre in Singapore and Brussels: European Policy Centre.

Garben, S. (2019). Competence Creep Revisited. *Journal of Common Market Studies, 57*(2), 205–222.

Garcia, M., & Masselot, A. (2015). The Value of Gender Equality in EU-Asian Trade Policy: An Assessment of the EU's Ability to Implement Its Own Legal Obligations. In A. Björkdahl, N. Chaban, J. Leslie, & A. Masselot (Eds.), *Importing EU Norms: Conceptual Framework and Empirical Findings* (pp. 191–210). Springer.

Gebhard, C. (2011). Coherence. In C. Hill & M. E. Smith (Eds.), *International Relations and the European Union* (2nd ed., pp. 101–127). Oxford University Press.

Gehring, M. W., & Robb, C. (2018). Sustainable Development and Emission Trading: The EU Perspective. In A. L. C. De Mestral, P. P. Fitzgerald, & T. Ahmad (Eds.), *Sustainable Development, International Aviation, and Treaty Implementation* (pp. 83–107). Cambridge University Press.

Gehring, T., Oberthür, S. & Mühleck, M. (2013). 'European Union Actorness in International Institutions: Why the EU is Recognized as an Actor in Some International Institutions, but Not in Others'. *Journal of Common Market Studies, 51* (5), 849–865.

Gippner, O. (2016). The 2 °C Target: A European Norm Enters the International Stage-Following the Process to Adoption in China. *International Environmental Agreements, 16*, 49–65.

Goron, C. (2014). EU–ASEAN Relations in the Post-2015 Climate Regime: Exploring Pathways for Top-down and Bottom-up Climate Governance. In W. Hofmeister & P. Rueppel (Eds.), *Climate Change Diplomacy: The Way Forward for Asia and Europe*, pp. 101–129. Kondrad Adenauer Stiftung; East Asian Institute; European Union Centre in Singapore and Brussels: European Policy Centre.

Groen, L. (2019). Explaining European Union Effectiveness (Goal Achievement) in the Convention on Biological Diversity: The Importance of Diplomatic Engagement. *International Environmental Agreements: Politics, Law and Economics, 19*(1), 69–87.

Groen, L., Niemann, A., & Oberthür, S. (2012). The EU as A Global Leader? The Copenhagen and Cancun UN Climate Change Negotiations. *Journal of Contemporary European Research, 8*(2), 173–191.

Groen, L., Niemann, A., & Oberthür, S. (2013). The EU Role in International Climate Change Policy-Making: A Global Leader in Decline? In A. Boening, J. F. Kremer, & Van Loon, A. (Eds.) *Global Power Europe—Vol. 2: Policies, Actions and Influence of the EU's External Relations* (pp. 37–54). Springer.

Guggenbühl, A. (2013). The Culture of Negotiation in the European Union: Reviewing Trends and Predicting Patterns of Multilateral Decision-Making. *The Hague Journal of Diplomacy, 8*(1), 21–47.

Gulbrandsen, C. (2013). Navigating from Conflict to Working Arrangement: EU Coordination in the International Maritime Organization. *Journal of European Integration, 35*(7), 749–765.

Gupta, J., & Ringius, L. (2001). The EU's Climate Leadership: Reconciling Ambition and Reality. *International Environmental Agreements: Politics, Law and Economics, 1*(2), 281–299.

Hackmann, B. (2012). Analysis of the Governance Architecture to Regulate GHG Emissions from International Shipping. *International Environmental Agreements: Politics, Law and Economics, 12*(1), 85–103.

Hadjiyianni, I. (2019). *The EU as A Global Regulator for Environmental Protection: A Legitimacy Perspective*. Hart Publishing.

Hadjiyianni, I., & Herodotou, A. (2023). The EU's Representation in International Organizations: Case C-161/20 *Commission v. Council (International Maritime Organization). European Foreign Affairs Review, 28*(1), 30–52.

Harrison, J. (2017). *Saving the Ocean Through Law: The International Legal Framework for the Protection of the Marine Environment*. Oxford University Press.

Haug, C., & Berkhout, F. (2010). Learning the Hard Way? European Climate Policy after Copenhagen. *Environment: Science and Policy for Sustainable Development, 52*(3), 20–27.

Hertogen, A. (2012). Sovereignty as Decisional Independence over Domestic Affairs: The Dispute over Aviation in the EU Emissions Trading System. *Transnational Environmental Law, 1*(2), 281–301.

Houghton, J. T., Jenkins, G. J., & Ephraums, J. J. (Eds.). (1990). *Climate Change: The IPCC Scientific Assessment*. Cambridge University Press.

Houghton, J. T., Ding, Y., Griggs, D. J., Noguer, M., van der Linden, P. J., Dai, X., Maskell, K., & Johnson, C. A. (Eds.). (2001). *Climate Change 2001: The Scientific Basis*. Cambridge University Press.

Houghton, J. T., Meira Filho, L. G., Callander, B. A., Harris, N., Kattenberg, A., & Maskell, K. (Eds.). (1995). *Climate Change 1995: The Science of Climate Change*. Cambridge University Press.

Huelss, H. (2012). A Force for Good Governance? The European Union's Normative Power and Standards of Appropriate Governing. *European Foreign Affairs Review, 17*(1), 93–112.

Hulme, M., & Mahony, M. (2010). Climate Change: What Do We Know about the IPCC? *Progress in Physical Geography, 34*(5), 705–718.

Hyde-Price, A. (2006). 'Normative' Power Europe: A Realist Critique. *Journal of European Public Policy, 13*(2), 217–234.

ICAO. (2013). *38th Session of the ICAO Assembly, 24 September to 4 October 2013, Montréal: List of Delegates No. 10.0.*

ICAO. (2016). *39th Session of the ICAO Assembly, 27 September to 7 October 2016, Montréal: List of Delegates as of 06-Oct-2016 16:33 hrs.*

ICAO. (2019a). *40th Session of the ICAO Assembly, 24 September to 4 October 2019, Montréal: List of Delegates as of 11-Oct-2019 16:16 hrs.*

ICAO. (2019b). *Project Final Report: Capacity Building for CO_2 Mitigation from International Aviation*. ICAO.

ICAO. (2022a). *About ICAO*. https://www.icao.int/about-icao/Pages/default.aspx. Last accessed: 25 January 2022.

ICAO. (2022b). *No Country Left Behind*. https://www.icao.int/about-icao/nclb/pages/default.aspx. Last accessed: 26 January 2022.

ICAO. (2022c, March). *Report on the Feasibility of a Long-term Aspirational Goal (LTAG) for International Civil Aviation CO_2 Emission Reductions*. https://www.icao.int/environmental-protection/LTAG/Documents/REPORT%20ON%20THE%20FEASIBILITY%20OF%20A%20LONG-TERM%20ASPIRATIONAL%20GOAL_en.pdf. Last accessed: 12 April 2022.

ICAO. (2022d). *The ICAO Council*. https://www.icao.int/about-icao/Council/Pages/Council.aspx. Last accessed: 26 January 2022.

ICAO. (2023). *Long Term Global Aspirational Goal (LTAG) for International Aviation*. https://www.icao.int/environmental-protection/Pages/LTAG.aspx. Last accessed: 8 August 2023.

Imber, M. F. (1989). *The USA, ILO, UNESCO and IAEA: Politicization and Withdrawal in the Specialized Agencies*. St. Martin's Press.

IMO. (2003, March 4). *Resolution A.963(23) Adopted on 5 December 2003 (Agenda Item 19)—IMO Policies and Practices Related to the Reduction of Greenhouse Gas Emissions from Ships*. A 23/Res.963.

IMO. (2015). *Third IMO Greenhouse Gas Study 2014*. International Maritime Organization.

IMO. (2017, February 21). *Consideration of How to Progress the Matter of Reduction of GHG Emissions from Ships: Existing IMO Activity Related to Reducing GHG Emissions in the Shipping Sector—Note by the Secerariat*. ISWG-GHG1/2.

IMO. (2018a). *Data Collection System for Fuel Oil Consumption of Ships*. http://www.imo.org/en/OurWork/Environment/PollutionPrevention/AirPollution/Pages/Data-Collection-System.aspx. Last accessed: 13 May 2019.

IMO. (2018b, April 16). *List of Participants*, MEPC 72/INF.1.

IMO (2018c, October 29). *List of Participants*, MEPC 73/INF.1.

IMO. (2019, May 21). *List of Participants*, MEPC 74/INF.1.

IMO. (2020). *Fourth IMO Greenhouse Gas Study 2020*. International Maritime Organization.

IMO. (2021, July 6). *List of Participants*, MEPC 76/INF.1.

IMO. (2022a). Energy Efficiency Measures. https://www.imo.org/en/OurWork/Environment/Pages/Technical-and-Operational-Measures.aspx. Last accessed: 20 March 2022.

IMO. (2022b). *Historic Background: IMO and the UNFCCC Framework*. https://www.imo.org/en/OurWork/Environment/Pages/Historic%20Background%20GHG.aspx. Last accessed: 19 March 2022.

IMO. (2022c). Structure of IMO. https://www.imo.org/en/About/Pages/Structure.aspx. Last accessed: 13 March 2022.

Ingold, K. (2011). Network Structures within Policy Processes: Coalitions, Power, and Brokerage in Swiss Climate Policy. *Policy Studies Journal, 39*(3), 435–459.

International Transport Forum. (2018). *Decarbonising Maritime Transport: Pathways to Zero-Carbon Shipping by 2035*. International Transport Forum Policy Papers, no. 47 OECD Publishing.

Jänicke, M., & Wurzel, R. K. W. (2019). Leadership and Lesson-drawing in the European Union's Multilevel Climate Governance. *Environmental Politics, 28*(1), 22–42.

Jordan, A., & Adelle, C. (Eds.). (2013). *Environmental Policy in the EU: Actors, Institutions and Processes* (3rd ed.). Routledge.

Jørgensen, K. E. (2009). The European Union in Multilateral Diplomacy. *The Hague Journal of Diplomacy, 4*(2), 189–200.

Jørgensen, K. E., & Laatikainen, K. V. (2004, 24–26 June). *The EU @ the UN: Multilateralism in a New Key?* Paper presented at the Second Pan-European Conference on EU Politics, Bologna.

Jørgensen, K. E., Oberthür, S., & Shahin, J. (2011). Introduction: Assessing the EU's Performance in International Institutions—Conceptual Framework and Core Findings. *Journal of European Integration, 33*(6), 599–620.

Juncker, J. C. (2014, November 1). Mission Letter: Violeta Bulc, Commissioner for Transport. *European Commission*. https://wayback.archive-it.org/12090/20191120070534/https://ec.europa.eu/commission/commissioners/sites/cwt/files/commissioner_mission_letters/bulc_en.pdf. Last accessed: 3 February 2022.

Juncos, A. (2007). *Cometh the 'Hour of Europe', Cometh the Institutions? Coherence and Effectiveness of the EU's Common Foreign and Security Policy in Bosnia and Herzegovina (1991–2006)* [PhD Thesis]. Loughborough: Loughborough University. https://dspace.lboro.ac.uk/2134/7895. 17 March 2019.

Juncos, A. E., & Pomorska, K. (2011). Invisible and Unaccountable? National Representative and Council Officials in EU Foreign Policy. *Journal of European Public Policy, 18*(8), 1096–1114.

Juncos, A., & Pomorska, K. (2020). The European External Action Service. *Oxford Research Encyclopedias: Politics*. https://doi.org/10.1093/acrefore/9780190228637.013.1081.

Kadelbach, S. (2019). ERTA Case. *Max Planck Encyclopedias of International Law*. Oxford: Oxford University Press.

Kassim, H., & Stevens, H. (2010). *Air Transport and the European Union: Europeanization and Its Limits*. Palgrave Macmillan.

Katsoulas, S. (2022). *The United States and Greek-Turkish Relations: The Guardian's Dilemma*. Routledge.

Katzenstein, P. J. (1996). Introduction: Alternative Perspectives on National Security. In P. J. Katzenstein (Ed.), *The Culture of National Security: Norms and Identity in World Politics* (pp. 1–32). Columbia University Press.

Kavalski, E. (2013). The Struggle for Recognition of Normative Power: Normative Power Europe and Normative Power China in Context. *Cooperation and Conflict, 48*(2), 247–267.

Keay Bright, S. (2001, April 6–11). *A Critical Analysis of the Voluntary Fuel Economy Agreements, Established between the Automobile Industry and the European Commission, with Regard for Their Capacity to Protect the Environment*. Paper presented at the ECPR Joint Sessions.

Keene, E. (2012). Social Status, Social Closure and the Idea of Europe as a 'Normative Power.' *European Journal of International Relations, 19*(4), 939–956.

Keukeleire, S. (2003). The European Union as a Diplomatic Actor: Internal, Traditional, and Structural Diplomacy. *Diplomacy and Statecraft, 14*(3), 31–56.

Keukeleire, S., Keuleers, F., & Raube, K. (2016). The EU, Structural Diplomacy and the Challenge of Learning. In M. Smith, S. Keukeleire, & S. Vanhoonacker (Eds.), *The Diplomatic System of the European Union: Evolution, Change and Challenges* (pp. 199–214). Routledge.

Keukeleire, S., Thiers, R., & Justaert, A. (2009). Reappraising Diplomacy: Structural Diplomacy and the Case of the European Union. *The Hague Journal of Diplomacy, 4*(2), 143–165.

Kissack, R. (2006). *Who Speaks for Europe in the ILO? Member State Coordination and European Union Representation in the International Labour Organization* [PhD Thesis]. London School of Economics and Political Sciences.

Kissack, R. (2007). *European Union Member State Coordination in the United Nations System: Towards a Methodology of Analysis*. European Foreign Policy Unit Working Paper 2007/1.

Klabbers, J. (2009). *An Introduction to International Institutional Law* (2nd ed.). Cambridge University Press.

Kleistra, Y., & Van Willigen, N. (2010). Evaluating Diplomacy: A Mission Impossible? *Evaluation, 16*(2), 119–135.

Kleistra, Y., & Van Willigen, N. (2015). Evaluating the Impact of EU Diplomacy: Pitfalls and Challenges. In J. A. Koops & G. Macaj (Eds.), *The European Union as a Diplomatic Actor* (pp. 52–68). Palgrave Macmillan.

Knill, C., & Liefferink, D. (2007). *Environmental Politics in the European Union: Policy-making, Implementation and Patterns of Multilevel Governance*. Manchester University Press.

Knill, C., & Tosun, J. (2009). Hierarchy, Networks, or Markets: How Does the EU Shape Environmental Policy Adoptions within and Beyond Its Borders? *Journal of European Public Policy, 16*(6), 873–894.

Kopela, S. (2017). Making Ships Cleaner: Reducing Air Pollution from International Shipping. *Review of European, Comparative & International Environmental Law, 26*(3), 231–242.

Korppoo, A. (2020). Domestic Frames on Russia's Role in International Climate Diplomacy. *Climate Policy, 20*(1), 109–123.

Kreppel, A., & Oztas, B. (2017). Leading the Band or Just Playing the Tune? Reassessing the Agenda-Setting Powers of the European Commission. *Comparative Political Studies, 50*(8), 1118–1150.

Kryżanowski, M. (2015). International Leadership Re-/constructed? Ambivalence and Heterogeneity of Identity Discourses in European Union's Policy on Climate Change. *Journal of Language and Politics, 14*(1), 110–133.

Kulovesi, K. (2012). Climate Change in EU External Relations: Please Follow My Example (Or I Might Force You To). In E. Morgera (Ed.), *The External Environmental Policy of the European Union: EU and International Law Perspectives* (pp. 115–148). Cambridge University Press.

Kulovesi, K., & Cremona, M. (2014). The Evolution of EU Competence in the Field of External Relations and its Impact on Environmental Governance Policies. In C. Bakker & F. Francioni (Eds.), *The EU, the US and Global Climate Governance* (pp. 81–94). Routledge.

Laatikainen, K. V., & Degrand Guillaud, A. (2010). Two Logics, One Treaty: The Lisbon Treaty and EU Foreign Policy in Brussels and at the UN. *Studia Diplomatica, 63*(1), 3–22.

Laïdi, Z. (Ed.). (2008). *EU Foreign Policy in a Globalized World: Normative Power and Social Preferences*. Routledge.

Larsen, H. (2014). The EU as a Normative Power and the Research on External Perceptions: The Missing Link. *Journal of Common Market Studies, 52*(4), 896–910.

Larsen, H. (2020). Normative Power Europe or Capability-expectations Gap? The Performativity of Concepts in the Study of European Foreign Policy. *Journal of Common Market Studies, 58*(4), 962–977.

Laursen, F. (2020). *The Development of the EU as A Sea-Policy Actor: Fish, Ships and Navies*. Edward Elgar Publishing.

Lavenex, S. (2014). The Power of Functionalist Extension: How EU Rules Travel. *Journal of European Public Policy, 21*(6), 886–903.

Lavenex, S., & Schimmelfenning, F. (2009). EU Rules Beyond EU Borders: Theorizing External Governance in European Politics. *Journal of European Public Policy, 16*(6), 791–812.

Lee, J. W. (2017). A New Global Approach to Mitigating Emissions from International Aviation: The International Civil Aviation Organization's 39th Assembly (2016). *Chinese Journal of Environmental Law, 1*(1), 104–110.

Leeg, T. (2014). Normative Power Europe? The European Union in the Negotiations on a Free Trade Agreement with India. *European Foreign Affairs Review, 19*(3), 335–356.

Lefevere, J., Runge Metzger, A., & Werksman, J. (2015). The EU and International Climate Change Policy. In J. Delbeke & P. Vis (Eds.), *EU Climate Policy Explained* (pp. 109–124). Routledge.

Lenk, H. (2015). Challenging the Notion of Coherence in EU Foreign Investment Policy. *European Journal of Legal Studies, 8*(2), 6–20.

Lewis, J. (2005). The Janus Face of Brussels: Socialization and Everyday Decision Making in the European Union. *International Organization, 59*(4), 937–971.

Light, A. (2017). Climate Diplomacy. In S. M. Gardiner & A. Thompson (Eds.), *The Oxford Handbook on Environmental Ethics* (pp. 487–500). Oxford University Press.

Lightfoot, S., & Burchell, J. (2005). The European Union and the World Summit on Sustainable Development: Normative Power Europe in Action? *Journal of Common Market Studies, 43*(1), 75–95.

Lin, W. (2017). Aviation and Climate Change: Practising Green Governmentality across the North-South Devide. *Geopolitics, 22*(1), 129–150.

Lindenthal, A. (2014). Aviation and Climate Protection: EU Leadership within the International Civil Aviation Organization. *Environmental Politics, 23*(6), 1064–1081.

Louis, M., & Maertens, L. (2021). *Why International Organizations Hate Politics: Depoliticizing the World*. Routledge.

Lütz, S., Leeg, T., Otto, D., & Woyames Dreher, V. (2021). *The European Union as A Global Actor: Trade, Finance and Climate Policy*. Springer.

Lyle, C. (2018). Beyond the ICAO's CORSIA: Towards a More Climatically Effective Strategy for Mitigation of Civil-Aviation Emissions. *Climate Law, 8*(1–2), 104–127.

Mabey, N., Gallagher, L., & Born, C. (2013). *Understanding Climate Diplomacy: Building Diplomatic Capacity and Systems to Avoid Dangerous Climate Change*. E3G Third Generation Environmentalism.

Macaj, G., & Nicolaïdis, K. (2014). Beyond 'one voice'? Global Europe's Engagement with Its Own Diversity. *Journal of European Public Policy, 21*(7), 1067–1083.

Maciej Kaczyński, P. (2014). EU Climate Diplomacy: Cautious Optimism Ahead of 2015. In W. Hofmeister & P. Rueppel (Eds.), *Climate Change Diplomacy: The Way Forward for Asia and Europe* (pp. 60–70). Kondrad Adenauer Stiftung; East Asian Institute; European Union Centre in Singapore and Brussels: European Policy Centre.

Magliveras, K. D. (2011). Membership in International Organizations. In J. Klabbers & Å. Wallendahl (Eds.), *Research Handbook on the Law of International Organizations* (pp. 84–107). Edward Elgar Publishing.

Malina, R., McConnachie, D., Wichester, N., Wollersheim, C., Paltsev, S., & Waitz, I. A. (2012). The Impact of the European Union Emissions Trading Scheme on US Aviation. *Journal of Air Transport Management, 19*, 36–41.

Manners, I. (2002). Normative Power Europe: A Contradiction in Terms? *Journal of Common Market Studies, 40*(2), 235–258.

Manners, I. (2006a). European Union 'Normative Power' and the Security Challenge. *European Security, 15*(4), 405–421.

Manners, I. (2006b). Normative Power Europe Reconsidered: Beyond the Crossroads. *Journal of European Public Policy, 13*(2), 182–199.

Manners, I. (2006c). The European Union as a Normative Power: A Response to Thomas Diez. *Millennium: Journal of International Studies, 35*(1), 167–180.

Manners, I. (2009a). The EU's Normative Power in Changing World Politics. In A. Gerrits (Ed.), *Normative Power Europe in A Changing World: A Discussion* (pp. 9–24). Netherlands Institute of International Relations Clingendael.

Manners, I. (2009b). The Social Dimension of EU Trade Policies: Reflection from a Normative Power Perspective. *European Foreign Affairs Review, 14*(5), 785–803.

Manners, I. (2010). Global Europa: Mythology of the European Union in World Politics. *Journal of Common Market Studies, 48*(1), 67–87.

Manners, I. (2013). Assessing the Decennial, Reassessing the Global: Understanding European Union Normative Power in Global Politics. *Cooperation and Conflict, 48*(2), 304–329.

Manners, I. (2015). Sociology of Knowledge and Production of Normative Power in the European Union's External Actions. *Journal of European Integration, 37*(2), 299–318.

Manners, I. (2021). Normative Power Approach to European Union External Action. In S. Gstöhl & S. Schunz (Eds.), *The External Action of the European Union: Concepts, Approaches, Theories* (pp. 69–84). Red Globe Press.

Manners, I., & Whitman, R. (2013). Normative Power and the Future of EU Public Diplomacy. In M. K. D. Cross & J. Melissen (Eds.), *European Public Diplomacy: Soft Power at Work* (pp. 183–204). Palgrave Macmillan.

Martinez Romera, B. (2018). *Regime Interaction and Climate Change: The Case of International Aviation and Maritime Transport*. Routledge.

Martinez Romera, B., & van Asselt, H. (2015). The International Regulation of Aviation Emissions: Putting Differential Treatment into Practice. *Journal of Environmental Law, 27*(2), 259–283.

Masson-Delmotte, V., Zhai, P., Pörtner, H. O., Roberts, D., Skea, J., Shukla, P. R., et al. (Eds.). (2018). *Global Warming of 1.5°C: An IPCC Special Report on the Impacts of Global Warming of 1.5°C Above Pre-Industrial Levels and Related Global Greenhouse Gas Emission Pathways, in the Context of Strengthening the Global Response to the Threat of Climate Change, Sustainable Development, and Efforts to Eradicate Poverty*. Intergovernmental Panel on Climate Change.

Mattocks, K. (2018). Co-ordinating Co-ordination: The European Commission and the Culture Open Method of Co-ordination. *Journal of Common Market Studies, 56*(2), 318–334.

Maurer, H. (2022). The Added Value of European Diplomacy for EU Regional and International Actorness. In M. R. Freire, P. D. Lopes, D. Nascimento, & L. Simão (Eds.), *EU Global Actorness in a World of Contested Leadership: Policies, Instruments and Perceptions* (pp. 35–52). Palgrave Macmillan.

Maurer, H., & Raik, K. (2018). Neither Fish nor Fowl. How EU Delegations Challenge the Institution of Diplomacy: The Cases of Moscow and Washington. *The Hague Journal of Diplomacy, 13*(1), 57–74.

Maurer, H., & Wright, N. (2020). A New Paradigm for EU Diplomacy? EU Council Negotiations in a Time of Physical Restrictions. *The Hague Journal of Diplomacy, 15*(4), 556–568.

Meunier, S. (2017). Integration by Stealth: How the European Union Gained Competence over Foreign Direct Investment. *Journal of Common Market Studies, 55*(3), 593–610.

Michaelis, P., & Zerle, P. (2006). From ACEA's Voluntary Agreement to an Emissions Trade Scheme for New Passenger Cars. *Journal of Environmental Planning and Management, 43*(3), 435–455.

Minas, S., & Ntousas, V. (Eds.). (2018). *EU Climate Diplomacy: Politics, Law and Negotiations*. Routledge.

Morgenstern-Pomorski, J. H. (2018). *The Contested Diplomacy of the European External Action Service: Inception, Establishment and Consolidation*. Routledge.

Morgera, E. (Ed.). (2012). *The External Environmental Policy of the European Union: EU and International Law Perspectives*. Cambridge University Press.

Musu, C. (2003). European Foreign Policy: A Collective Policy or a Policy of 'Converging Parallels? *European Foreign Affairs Review, 8*(1), 35–49.

Nekvasil, M., & Moldan, B. (2018). Could Cohesion Policy Push EU Climate Efforts? *Climate Policy, 18*(1), 129–139.

Nengye, L., & Maes, F. (2010). The European Union and the International Maritime Organization: EU's External Influence on the Prevention of Vessel-Source Pollution. *Journal of Maritime Law & Commerce, 41*(4), 581–594.

Nengye, L., & Maes, F. (2012). Legal Constraints to the European Union's Accession to the International Maritime Organization. *Journal of Maritime Law & Commerce, 43*(2), 279–291.

Neuman, M. (Ed.). (2019). *Democracy Promotion and the Normative Power Europe Framework: The European Union in South Eastern Europe, Eastern Europe, and Central Asia*. Springer.

Newburger, E. (2020, October 24). Joe Biden Calls Climate Change the 'Number One Issue Facing Humanity'. *CNBC*. https://www.cnbc.com/2020/10/24/joe-biden-climate-change-is-number-one-issue-facing-humanity.html. Last accessed: 31 May 2022.

Noutcheva, G. (2008). Enlargement Policy towards Central and Eastern Europe, 1989–2007: Normative Intended. In N. Tocci (Ed.), *Who Is A Normative Foreign Policy Actor? The European Union and Its Global Partners* (pp. 26–31). Centre for European Policy Studies.

Noutcheva, G. (2009). Fake, Partial and Imposed Compliance: The Limits of the EU's Normative Power in the Western Balkans. *Journal of European Public Policy, 16*(7), 1065–1084.

Nowak, B. (2014). Domestic Sources of Multilateral Diplomacy—The Case of Poland's Climate Policy. In W. Hofmeister & P. Rueppel (Eds.), *Climate Change Diplomacy: The Way Forward for Asia and Europe* (pp. 53–59). Kondrad Adenauer Stiftung; East Asian Institute; European Union Centre in Singapore and Brussels: European Policy Centre.

Oberthür, S. (2003). Institutional Interaction to Address Greenhouse Gas Emissions from International Transport: ICAO, IMO and the Kyoto Protocol. *Climate Policy, 3*(3), 191–205.

Oberthür, S. (2006). The Climate Change Regime: Interactions with ICAO, IMO, and the EU Burden-Sharing Agreement. In S. Oberthür & T. Gehring (Eds.), *Institutional Interaction in Global Environmental Governance Synergy and Conflict among International and EU Policies* (pp. 53–77). MIT Press.

Oberthür, S. (2011). The European Union's Performance in the International Climate Change Regime. *Journal of European Integration, 33*(6), 667–682.

Oberthür, S. (2016). Where to Go from Paris? The European Union in Climate Geopolitics. *Global Affairs, 2*(2), 119–130.

Oberthür, S., & Dupont, C. (2021). The European Union's International Climate Leadership: Towards A Grand Climate Strategy? *Journal of European Public Policy, 28*(7), 1095–1114.

Oberthür, S., & Groen, L. (2014). EU Performance in the International Climate Negotiations in 2013: Scope for Improvement. *Institute for European Studies*, Policy brief, iss. 01.

Oberthür, S., & Groen, L. (2015). The Effectiveness Dimension of the EU's Performance in International Institutions: Toward A More Comprehensive Assessment Framework. *Journal of Common Market Studies, 53*(6), 1319–1335.

Oberthür, S., & Groen, L. (2018). Explaining Goal Achievement in International Negotiations: The EU and the Paris Agreement on Climate Change. *Journal of European Public Policy, 25*(5), 708–727.

Oberthür, S., & Pallemaerts, M. (2010a). The EU's Internal and External Climate Policies: An Historical Overview. In S. Oberthür & M. Pallemaerts (Eds.), *The New Climate Policies of the European Union: Internal Legislation and Climate Diplomacy* (pp. 27–64). Brussels University Press.

Oberthür, S., & Pallemaerts, M. (Eds.). (2010b). *The New Climate Policies of the European Union: Internal Legislation and Climate Diplomacy*. Brussels University Press.

Oberthür, S., & Rabitz, F. (2014). On the EU's Performance and Leadership in Global Environmental Governance: The Case of the Nagoya Protocol. *Journal of European Public Policy, 21*(1), 39–57.

Oberthür, S., & Roche Kelly, C. (2008). EU Leadership in International Climate Policy: Achievements and Challenges. *The International Spectator, 43*(3), 35–50.

Official Journal of the European Commission. (2001). *Commission: European Governance—A White Paper COM(2001) 428 final*, C 278/1.

Official Journal of the European Union. (2003). *Directive 2003/87/EC of the European Parliament and of the Council of 13 October 2003 Establishing a Scheme for Greenhouse Gas Emissions Allowance Trading Within the Community and Amending Council Directive 96/61/EC (Text with EEA relevance)*, L 275/32.

Official Journal of the European Union. (2009). *Directive 2008/101/EC of the European Parliament and the Council of 19 November 2008 Amending Directive 2003/87/EC so as to Include Aviation Activities in the Scheme for Greenhouse Gas Emission Allowance Trading Within the Community (Text with EEA Relevance)*, L 8/3.

Official Journal of the European Union. (2011). *Council Decision of 31 March 2011 on the Signing, on Behalf of the Union, and Provisional Application of a Memorandum of Cooperation between the European Union and the International Civil Aviation Organization Providing a Framework for Enhanced Cooperation (2011/530/EU)*, L 232/1.

Official Journal of the European Union. (2015). *Regulation (EU) 2015/757 of the European Parliament and of the Council of 29 April 2015 on the Monitoring, Reporting and Verification of Carbon Dioxide Emissions from Maritime Transport, and Amending Directive 2009/16/EC (Text with EEA relevance)*, L 123/55.

Official Journal of the European Union. (2016a). *Council Decision (EU) 2016/915 of 30 May 2016 on the position to be taken on behalf of the European Union with Regard to the International Instrument to Be Drawn Up within the ICAO Bodies and Intended to Lead to the Implementation from 2020 of A Single Global-Based Measure for International Aviation Emissions*, L 153/32.

Official Journal of the European Union. (2016b). *Directive (EU) 2016/802 of the European Parliament and of the Council of 11 May 2016 relating to a reduction in the sulphur content of certain liquid fuels (codification)*, L 132/58.

Official Journal of the European Union. (2018). *Directive (EU) 2018/410 of the European Parliament and of the Council of 14 March 2018 amending Directive 2003/87/EC to enhance cost-effective emission reductions and low-carbon investments, and Decision (EU) 2015/1814 (Text with EEA relevance)*, L 76/3.

Official Journal of the European Union. (2021). *Regulation (EU) 2021/1119 of the European Parliament and of the Council of 30 June 2021 establishing the framework for achieving climate neutrality and amending Regulations (EC) No 401/2009 and (EU) 2018/1999 ('European Climate Law)'*, L 243/1.

Ott, H. E. (2001). Climate Change: An Important Foreign Policy Issue. *International Affairs, 77*(2), 277–296.

Pace, M. (2007). The Construction of EU Normative Power. *Journal of Common Market Studies, 45*(5), 1041–1064.

Pallis, A. A. (2002). *The Common EU Maritime Transport Policy: Policy Europeanisation in the 1990s*. Routledge.

Panke, D. (2014). The European Union in the United Nations: An Effective External Actor? *Journal of European Public Policy, 21*(7), 1050–1066.

Panke, D. (2017). Speaking with One Voice: Easier Said than Done? The EU in the UNGA. In S. Blavoukos & D. Bourantonis (Eds.), *The EU in UN Politics: Actors, Processes and Performances* (pp. 27–46). Palgrave Macmillan.

Parker, C. F., & Karlsson, S. (2017). The European Union as A Global Climate Leader: Confronting Aspiration with Evidence. *International Environmental Agreements: Politics, Law and Economics, 17*(4), 445–461.

Parker, C. F., Karlsson, S., & Hjerpe, M. (2017). Assessing the European Union's Global Climate Change Leadership: From Copenhagen to the Paris Agreement. *Journal of European Integration, 39*(2), 239–252.

Parker, O., & Rosamond, B. (2013). NPE Meets Economic Liberalism: Complicating Cosmopolitanism Inside/Outside the EU. *Cooperation and Conflict, 48*(2), 229–246.

Parry, M., Canziani, O., Palutikof, J., van der Linden, P., & Hanson, C. (Eds.). (2007). *Climate Change 2007: Impacts, Adaptation and Vulnerability.* Cambridge University Press.

Penner, J. E., Lister, D. H., Griggs, D. J., Dokken, D. J., & McFarland, M. (Eds.). (1999). *Summary for Policy Makers: Aviation and the Global Atmosphere*. Intergovernmental Panel on Climate Change.

Petersen, M. (2008). The Legality of the EU's Stand-Alone Approach to the Climate Impact of Aviation: The Express Role Given to the ICAO by the Kyoto Protocol. *Review of European, Comparative & International Environmental Law, 17*(2), 196–204.

Petrov, P., Pomorska, K., & Vanhoonacker, S. (2012). The Emerging EU Diplomatic System: Opportunities and Challenges after Lisbon. *The Hague Journal of Diplomacy, 7*(1), 1–9.

POLITICO Europe. (2020). *Opening interview with EU Commissioner Vălean | POLITICO's Sustainable Future Summit*. https://www.youtube.com/watch?v=Gp1_HW1ebyk. Last accessed: 9 February 2022.

Pomorska, K., & Vanhoonacker, S. (2016). 'Europe as a Global Actor: Searching for a New Strategic Approach. *Journal of Common Market Studies, 54*(Annual Review), 204–217.

Pomorska, K., & Wright, N. (2013). Europeanization and the Common and Security Policy. In C. Bretherton & M. Mannin (Eds.), *The Europeanization of European Politics* (pp. 151–162). Palgrave Macmillan.

Ponzano, P., Hermanin, C., & Corona, D. (2012). The Power of Initiative of the European Commission: A Progressive Erosion? *Notre Europe*, Study & Research 89.

Pregernig, M., & Böcher, M. (2012). Normative and Analytical Perspective on the Role of Science and Expertise in Environmental Governance. In K. Hogl, E. Kvarda, R. Nordbeck, & M. Pregernig (Eds.), *Environmental Governance: The Challenges of Legitimacy and Effectiveness* (pp. 199–219). Edward Elgar Publishing.

Preston, H., Lee, D. S., & Hooper, P. D. (2012). The Inclusion of the Aviation Sector within the European Union's Emissions Trading Scheme: What Are the Prospects for a More Sustainable Aviation Industry? *Environmental Development, 2*, 48–56.

Prum, D. A., & Kisska-Schulze, K. (2015). The Environmentally Conscious Skies: Did the European Union's Game of Brinksmanship Lead to a Viable

Global Plan for Emissions Trading in Aviation? *Washington University Global Studies Review, 14*(1), 1–48.

Rama Montaldo, M. (1970). International Legal Personality and Implied Powers of International Organizations. *British Yearbook of International Law, 44*, 111–155.

Rasch, M. (2008). *The European Union and the United Nations: The Functioning and Coherence of EU External Representation in a State-centric Environment*. Martinus Nijhoff Publishers.

Rathbun, B. C. (2008). Interviewing and Qualitative Field Methods: Pragmatism and Practicalities. In J. M. Box Steffensmeier, H. E. Brady, & D. Collier (Eds.), *The Oxford Handbook of Political Methodology* (pp. 685–701). Oxford University Press.

Rayner, T., & Jordan, A. (2016). Climate Change Policy in the European Union. *Oxford Research Encyclopedia: Climate Science*. https://doi.org/10.1093/acrefore/9780190228620.013.47

Reagan, D. B. (2008). Putting International Aviation into the European Union Emissions Trading Scheme: Can Europe Do It Flying Solo? *Boston College Environmental Affairs Law, 35*(2), 349–384.

Renckens, S., Skogstad, G., & Mondu, M. (2017). When Normative and Market Power Interact: The European Union and Global Biofuels Governance. *Journal of Common Market Studies, 55*(6), 1432–1448.

Riddervold, M. (2016). (Not) in the Hands of the Member States: How the European Commission Influences EU Security and Defence Policies. *Journal of Common Market Studies, 54*(2), 353–369.

Riddervold, M., & Rosén, G. (2015). Beyond Intergovernmental Cooperation: The Influence of the European Parliament and the Commission on EU Foreign and Security Policies. *European Foreign Affairs Review, 20*(3), 399–418.

Riddervold, M., & Rosén, G. (2016). Trick and Treat: How the Commission and the European Parliament Exert Influence in European Foreign and Security Policies. *Journal of European Integration, 38*(6), 687–702.

Rietig, K. (2019). The Importance of Compatible Beliefs for Effective Climate Policy Integration. *Environmental Politics, 28*(2), 228–247.

Ritchie, H. (2020, October 22). Climate Change and Flying: What Share of Global CO_2 Emissions Come from Aviation? *Our World Data*. https://ourworldindata.org/co2-emissions-from-aviation. Last accessed: 16 March 2022.

Roccu, R., & Voltolini, B. (2018). Framing and Reframing the EU's Engagement with Mediterranean: Examining the Security-Stability Nexus Before and After the Arab Uprisings. *Mediterranean Politics, 23*(1), 1–22.

Romanyshyn, I. (2015). Explaining EU Effectiveness in Multilateral Institutions: The Case of the Arms Trade Treaty Negotiations. *Journal of Common Market Studies, 53*(4), 875–892.

Rony, A. H., Kitada, M., Dalaklis, D., Ölçer, A. I., & Ballini, F. (2019). Exploring the New Policy Framework of Environmental Performance Management for Shipping: A Pilot Study. *WMU Journal of Maritime Affairs, 18*(1), 1–24.

Russo, E. (2017). Towards an Exclusive Competence of the EU to Conclude Climate Agreements? *European Foreign Affairs Review, 22*(2), 197–212.

Scheipers, S., & Sicurelli, D. (2007). Normative Power Europe: A Credible Utopia? *Journal of Common Market Studies, 45*(2), 435–457.

Schimmelfennig, F., & Thomas, D. C. (2009). Normative Institutionalism and EU Foreign Policy in Comparative Perspective. *International Politics, 46*(4), 491–504.

Schmitz, L., & Eimer, T. R. (2020). From Coherence to Coheritization: Explaining the Rise of Policy Coherence in EU External Policy. *Globalizations, 17*(4), 629–647.

Schout, A., & Jordan, A. (2005). Coordinated European Governance: Self-organizing or Centrally Steered? *Public Administration, 83*(1), 201–220.

Schunz, S. (2009). The European Union's External Climate Policy: A Foreign Policy Analysis. *CFSP Forum, 7*(2), 6.

Schunz, S. (2014). *European Foreign Policy and the Global Climate Regime.* P.I.E. Peter Lang.

Schunz, S. (2015). The European Union's Climate Change Diplomacy. In J. A. Koops & G. Macaj (Eds.), *The European Union as A Diplomatic Actor* (pp. 178–201). Palgrave Macmillan.

Schunz, S., Basu, S., Bruyninckx, H., Keukeleire, S. & Wouters, J. (2012). 'Analysing the Position of the European Union in the United Nations System: Analytical Framework'. In Wouters, J., Bruyninckx, H., Basu, S. & Schunz, S. (Eds.)., *The European Union and Multilateral Governance: Assessing EU Participation in the United Nations Human Rights and Environmental Fora*(Basingstoke: Palgrave Macmillan), 25–48.

Schunz, S. & Damro, C. (2020). 'Expanding Actorness to Explain EU External Engagement in Originally Internal Policy Areas'. *Journal of European Public Policy, 27* (1), 122–140.

Schunz, S., Happaerts, S., & Van Den Brande, K. (2009). European Union Foreign Policy and Global Climate Change: Towards A Comprehensive European Climate Diplomacy?. *Leuven Centre for Global Governance Studies*, Policy Brief No. 12.

Scott, J., & Rajamani, L. (2012). EU Climate Change Unilateralism. *European Journal of International Law, 23*(2), 469–494.

Shi, Y. (2018). The Implications of the Paris Agreement for the Regulation of Greenhouse Gas Emissions from International Shipping. *Ocean Yearbook, 32*, 528–555.

Shi, Y., & Gullet, W. (2018). International Regulation on Low-Carbon Shipping for Climate Change Mitigation: Development, Challenges and Prospects. *Ocean Development & International Law, 49*(2), 134–156.

Sida's Helpdesk for Environment and Climate Change. (2012). *Western Balkan—Environment and Climate Change Policy Brief*. University of Gothenburg & SLU.

Sjöstedt, G. (1998). The EU Negotiates Climate Change: External Performance and Internal Structural Change. *Cooperation and Conflict, 33*(3), 227–256.

Sjursen, H. (2015). Normative Theory: An Untapped Resource in the Study of European Foreign Policy. In K. E. Jørgensen, Å. Kalland Aarstad, E. Drieskens, K. Laatikainen, & B. Tonra (Eds.), *The SAGE Handbook on European Foreign Policy* (Vol. 1, pp. 196–214). Sage.

Sjursen, H. (2006a). The EU as a 'Normative' Power: How Can This Be? *Journal of European Public Policy, 13*(2), 235–251.

Sjursen, H. (2006b). What Kind of Power? *Journal of European Public Policy, 13*(2), 169–181.

Skjærseth, J. B., & Wettestad, J. (2008a). *EU Emissions Trading: Initiation, Decision-making and Implementation*. Ashgate.

Skjærseth, J. B., & Wettestad, J. (2008b). Implementing EU Emissions Trading: Success or Failure? *International Environmental Agreements: Politics, Law and Economics, 8*(3), 275–290.

Smith, K. E. (2006). Speaking with One Voice? European Union Co-ordination on Human Rights Issues at the United Nations. *Journal of Common Market Studies, 44*(1), 113–137.

Sochor, E. (1991). *The Politics of International Aviation*. Macmillan Press.

Spence, D., & Bátora, J. (Eds.). (2015). *The European External Action Service: European Diplomacy Post-Westphalia*. Palgrave Macmillan.

Stevens, H. (2004). *Transport Policy in the European Union*. Palgrave Macmillan.

Stevens, H. (2021, September 6–8). *The Management of Mature Policies—The Case of Transport*. Paper presented at the UACES 51st Annual Conference.

Tänzler, D., & Carius, A. (2013). Beyond International Climate Negotiations: Climate Diplomacy from a Foreign Policy Perspective. In O. C. Ruppel, C. Roschmann, & K. Ruppel Schlichting (Eds.), *Climate Change: International Law and Global Governance, Volume II: Policy, Diplomacy and Governance in A Changing Environment* (pp. 259–274). Nomos.

Terpan, F. (2015). Soft Law in the European Union—The Changing Nature of EU Law. *European Law Journal, 21*(1), 68–96.

The Business Times. (2018, August 2). *EU an Avid Supporter of Rules-Based Global System*, Singapore. https://www.businesstimes.com.sg/opinion/eu-an-avid-supporter-of-rules-based-global-system. Last accessed: 13 March 2022.

The Guardian. (2021, October 14). UN Chief Urges Airlines and Shipping Firms to Do More to Cut Emissions. https://www.theguardian.com/environment/2021/oct/14/un-chief-urges-airlines-and-shipping-firms-to-do-more-to-cut-emissions?fbclid=IwAR2wlclWt2FajfrdrVHqj9OHDATumxNzZyAcyfcYYDfpav5MJYmCo2CnnAA. Last accessed: 13 March 2022.

Thomas, D. C. (2009). Explaining the Negotiation of EU Foreign Policy: Normative Institutionalism and Alternative Explanations. *International Politics, 46*(4), 339–357.

Thomas, D. C. (2012). Still Punching below Its Weight? Coherence and Effectiveness in European Union Foreign Policy. *Journal of Common Market Studies, 50*(3), 457–474.

Thomas, D. C., & Tonra, B. (2012). To What Ends EU Foreign Policy? Contending Approaches to the Union's Diplomatic Objectives and Representation. *The Hague Journal of Diplomacy, 7*(1), 11–29.

Tocci, N. (2007). *Profiling Normative Foreign Policy: The European Union and Its Global Partners*. CEPS Working Document, no. 279.

Tocci, N. (Ed.). (2008). *Who Is A Normative Foreign Policy Actor? The European Union and Its Global Partners*. Centre for European Policy Studies.

Torney, D. (2014). External Perceptions and EU Foreign Policy Effectiveness: The Case of Climate Change. *Journal of Common Market Studies, 52*(6), 1358–1373.

Torney, D. (2019). Follow the Leader? Conceptualising the Relationship between Leaders and Followers in Polycentric Climate Governance. *Environmental Politics, 28*(1), 167–186.

Traut, M., Larkin, A., Anderson, K., McGlade, C., Sharmina, M., & Smith, T. (2018). CO_2 Abatement Goals for International Shipping. *Climate Policy, 18*(8), 1066–1075.

United Nations. (1992). *United Nations Framework Convention on Climate Change*.

Van der Mei, A. P. (2016). Case Note: EU External Relations and Internal Inter-Institutional Conflicts: The Battlefield of Article 218 TFEU. *Maastricht Journal of European and Comparative Law, 23*(6), 1051–1076.

Van Evera, S. (1997). *Guide to Methods for Students of Political Science*. Cornell University Press.

Van Leeuwen, J. (2015). The Regionalization of Maritime Governance: Towards a Polycentric Governance System for Sustainable Shipping in the European Union. *Ocean & Coastal Management, 117*, 23–31.

Van Leeuwen, J. (2019). Capturing a Moving Target: Decarbonizing Shipping Through Informational Governance. In P. G. Harris (Ed.), *Climate Change and Ocean Governance: Politics and Policy for Threatened Seas* (pp. 376–392). Cambridge University Press.

Van Leeuwen, J., & Kern, K. (2013). The External Dimension of European Union Marine Governance: Institutional Interplay between the EU and the International Maritime Organization. *Global Environmental Politics, 13*(1), 69–87.

Van Schaik, L. (2010). The Sustainability of the EU's Model for Climate Diplomacy. In S. Oberthür & M. Pallemaerts (Eds.), *New Climate Policies of the European Union: Internal Legislation and Climate Diplomacy* (pp. 251–280). Brussels University Press.

Van Schaik, L., & Schunz, S. (2012). Explaining EU Activism and Impact in Global Climate Politics: Is the Union a Norm- or Interest-Driven Actor? *Journal of Common Market Studies, 50*(1), 169–186.

Van Willigen, N., & Kleistra, Y. (2013). Evaluating Diplomacy. In K. E. Jørgensen & K. V. Laatikainen (Eds.), *Routledge Handbook on the European Union and International Institutions: Performance, Policy, Power* (pp. 102–112). Routledge.

Vanhoonacker, S., & Pomorska, K. (2013). The European External Action Service and Agenda-setting in European Foreign Policy. *Journal of European Public Policy, 20*(9), 1316–1331.

Vaugeois, S. (2019). International Civil Aviation Organization (ICAO): The Developments of the Position of the EU in International Aviation. In R. A. Wessel & J. Odermatt (Eds.), *Research Handbook on the European Union and International Organizations* (pp. 222–239). Edward Elgar Publishing.

Vedder, H. (2012). The Formalities and Substance of EU External Environmental Competence: Stuck between Climate Change and Competitiveness. In E. Morgera (Ed.), *The External Environmental Policy of the European Union: EU and International Law Perspectives* (pp. 11–32). Cambridge University Press.

Veno, J. (2007). Flying the Unfriendly Skies: The European Union's New Proposal to Include Aviation in their Emissions Trading Scheme. *Journal of Air Law and Commerce, 27*(3), 659–687.

Vogler, J. (2009). Climate Change and EU Foreign Policy: The Negotiation of Burden Sharing. *International Politics, 46*(4), 469–490.

Vogler, J. (2013). Energy and Climate Policy. In C. Bretherton & M. Mannin (Eds.), *The Europeanization of European Politics* (pp. 137–149). Palgrave Macmillan.

Vogler, J. (2014). EU Climate Diplomacy—Internal and External Dimensions. In W. Hofmeister & P. Rueppel (Eds.), *Climate Change Diplomacy: The Way Forward for Asia and Europe* (pp. 71–88). Konrad Adenauer Stiftung; East Asian Institute; European Union Centre in Singapore and Brussels: European Policy Centre.

von der Leyen, U. (2019a). *A Union That Strives for More—My Agenda for Europe: Political Guidelines for the Next European Commission 2019–2024*. https://ec.europa.eu/info/sites/default/files/political-guidelines-next-commission_en_0.pdf. Last accessed: 3 February 2022.

von der Leyen, U. (2019b, December 1). *Mission Letter: Adina Vălean, Commissioner for Transport*. https://ec.europa.eu/commission/commissioners/sites/default/files/commissioner_mission_letters/mission-letter-valean-2019-2024_en.pdf. Last accessed: 3 February 2022.

Watanabe, R., & Robinson, G. (2005). The European Union Emissions Trading Scheme (EU ETS). *Climate Policy, 5*(1), 10–14.

Weber, L. (2021). *International Civil Aviation Organization (ICAO)* (4th ed.). Wolters Kluwer.

Wettestad, J., Eikeland, P. O., & Nilsson, M. (2012). EU Climate and Energy Policy: A Hesitant Supranational Turn? *Global Environmental Politics, 12*(2), 67–86.

Whichmann, N. (2010). *Rule of Law Promotion in the European Neighbourhood Policy: Normative or Strategic Power?* Nomos.

Whichmann, N. (2007). Promoting the Rule of Law in the European Neighbourhood Policy—Strategic or Normative Power? *Politique Européenne, 2*(22), 81–104.

White, P. (1999). The Role of UN Specialised Agencies in Complex Emergencies: A Case Study of FAO. *Third World Quarterly, 20*(1), 223–238.

Whitelaw, K. (2009, December 19). U.S., China Reach Tentative Climate Compromise. *NPR*. https://www.npr.org/templates/story/story.php?storyId=121600755&t=1649206689992. Last accessed: 31 May 2022.

Whiteman, A. (2018, May 14). IMO Listens to Members of European Parliament to Act to Reduce Shipping Emissions. *Canadian Sailings*, p. 48.

Whitman, R. (2013). The Neo-Normative Turn in Theorizing the EU's International Presence. *Cooperation and Conflict, 48*(2), 171–193.

Wood, S. (2009). The European Union: A Normative or Normal Power? *European Foreign Affairs Review, 14*(2), 235–258.

Wouters, J., de Jong, S., Marx, A., & De Man, P. (2009). *Study for the Assessment of the EU's Role in International Maritime Organizations*. Leuven Centre for Global Governance Studies, Katholieke Universiteit Leuven.

Wurzel, R. K. W., & Connelly, J. (2011). Introduction: European Union Political Leadership in International Climate Change Politics. In R. K. W. Wurzel & J. Connelly (Eds.), *The European Union as a Leader in International Climate Change Politics* (pp. 3–20). Routledge.

Youngs, R. (2004). Normative Dynamics and Strategic Interests in the EU's External Identity. *Journal of Common Market Studies, 42*(2), 415–435.

Zervaki, A. (2014). *Resetting the Political Culture Agenda: From Polis to International Organization*. Springer.

Zhang, H. (2016). Towards Global Green Shipping: The Development of International Regulation on Reduction of GHG Emissions from Ships. *International Environmental Agreements: Politics, Law and Economics, 16*(4), 561–577.

GPSR Compliance
The European Union's (EU) General Product Safety Regulation (GPSR) is a set of rules that requires consumer products to be safe and our obligations to ensure this.

If you have any concerns about our products, you can contact us on

ProductSafety@springernature.com

In case Publisher is established outside the EU, the EU authorized representative is:

Springer Nature Customer Service Center GmbH
Europaplatz 3
69115 Heidelberg, Germany

www.ingramcontent.com/pod-product-compliance
Ingram Content Group UK Ltd.
Pitfield, Milton Keynes, MK11 3LW, UK
UKHW020835290926
15470UKWH00004B/1360
* 9 7 8 3 0 3 1 5 1 1 2 2 6 *